DIAGNOSIS AND PRESCRIPTION IN THE CLASSROOM:
Some Common Maths Problems

DIAGNOSIS AND PRESCRIPTION IN THE CLASSROOM:

Some Common Maths Problems

Ruth Rees and George Barr

Harper & Row, Publishers
London

Cambridge
Hagerstown
Philadelphia
New York

San Francisco
Mexico City
São Paulo
Sydney

Copyright © 1984 Ruth Rees and George Barr
All rights reserved
First published 1984
Harper & Row Ltd
28 Tavistock Street
London WC2E 7PN

No part of this book may be reproduced in any manner whatsoever without written permission except in the case of brief quotations in critical articles or reviews.

Rees, Ruth
  Diagnosis and prescription in the
  classroom.
  1. Mathematics—1961 -
  I. Title II. Barr, George 1956 -
  510 QA39.2

ISBN 0-06-318269-6

Typeset & Illustrated by Burns & Smith, Derby.
Printed & bound by Butler & Tanner Ltd, Frome and London.

The Mathematics Education Group at Brunel

This book has arisen from the activities of our team co-ordinated through MEG. Throughout the writing, Colin Winter has actively engaged with us in discussion of the text in addition to his specific contribution of Chapter 12. Jim Curnyn has remained in close contact with us since he left the group and Chapter 10 has been based on his work. Since the setting up of MEG we have much appreciated the guidance of Professor Desmond Furneaux and his vision and insight.

Ruth Rees
George Barr

Professor Desmond Furneaux embarked on a career as a physicist at what is now the G.E.C. Hirst Centre, before developing an interest in psychology which eventually took him to the University of London's Institute of Psychiatry where he spent 15 years on research which included work on human intellectual functions. He moved then to the Inner London Education Authority, to set up their Research and Statistics Group, before finally settling at Brunel in 1966 as Professor and Head in its newly established Education Department. He resigned the headship in 1981, but continues part-time as professor and director of PEDOG, the department's externally funded research organization, within which the MEG functions.

Colin F. Winter Senior Lecturer in Education (mathematics), is concerned with teacher training and research in mathematics education. Taught mathematics, statistics and computing in secondary schools and on part-time courses with adults. Prior to joining Brunel in 1968 was Head of Mathematics in a large London grammar school.

Ruth Rees Senior Research Fellow and Lecturer in Education (mathematics); as such initiated and directs the research activities of MEG and is involved with teacher training. Prior to joining Brunel in 1970 had considerable experience in teaching mathematics and physics in secondary and teriary education as well as some primary-school experience. Posts included appointments as head of Mathematics and Science Departments in schools.

George Barr Research Fellow in mathematics education. Taught mathematics and statistics in institutes of further and higher education. Joined MEG in 1978 to research the difficulties students of all ages experience in learning mathematics.

James C. Curnyn Educational Psychologist. Taught mathematics and statistics in secondary schools. Joined MEG in 1977 for 3 years to research the development of attitudes, anxiety and mathematical concepts for school pupils at late primary and early secondary levels.

ACKNOWLEDGEMENTS

Many people have been involved in the preparation of this book. Our thanks are due in large measure to all those teachers and their students who have so cheerfully worked with us in our studies. We were most encouraged in the early stages of this work to be told even by some of the tougher further-education students 'Anyone who can help us with our maths is doing a worthwhile job'! And it goes without saying that we have greatly appreciated the time which many teachers have devoted as co-workers.

The studies could not have taken place without the considerable financial support provided by many funding bodies such as DES, the Nuffield Foundation, the Leverhulme Trust and more recently CGLI and MSC. In particular we wish to thank British Petroleum Company plc not only for their generous provision of a 5-year research fellowship for one of us but for the considerable involvement and interest of Mr Peter Coaker and Mr Jim Ball.

The text incorporates data from many different kinds of studies: interviews carried out for the BBC series *Maths with Meaning* and our student teachers' projects such as that of Ms Sharon Slade with pre-TOPS adults.

We have much appreciated the very helpful comments of reviewers in the early stages of preparation of the manuscript and also those of Mr L. J. Oldknow who read through the final version as a 'layman'. We have taken on board many of the suggestions given.

Our thanks are especially due to our secretaries Mrs Renate Feder and Mrs Pamela Wells who so patiently typed and retyped the manuscript. Finally (we hope we have not left anyone out) to Marianne Lagrange, our editor, goes our appreciation for keeping us firmly but diplomatically to our deadlines!

PREFACE

In 1970, Ruth Rees came flying on a shoe-string, into the department of education at Brunel University — that is to say, we couldn't really afford her! Our DES funded Further Education research group was already working to the limits of its budget, but a problem had been unearthed which seemed too important to ignore. Ruth's job was to help us decide whether it really was as important as it seemed.

The FE group members were studying a variety of topics. Why did such a large proportion of college students drop out of their courses without even attempting examinations? Why were the failure rates in so many examinations so high? How could selection for advanced FE courses be improved? Why were so many FE teachers disappointed (and bewildered) by the results their students achieved? Why were so many employers so critical of the quality of the education their apprentices seemed to have received?

Researchers are accustomed to find that attempts to answer their questions simply direct attention to fresh questions. We were therefore intrigued to find that many experienced teachers in FE (particularly those working with craft, technician and ONC students) believed that many of our questions could receive a simple direct answer, and that this answer was much the same for all of them — 'these kids can't do calculations. And for most of them, we don't seem to be able to do much about it.' Of course, it was not exactly news, that mathematics was not everyone's favourite subject, and systematic study[1] of the attitudes and opinions of craft and technician students had already directed attention to their uncertainties and anxieties about maths, and to their dissatisfaction with the quality of the instruction they received in that subject. But the idea that difficulties with a particular limited area within

---

[1] e.g. Moore B.M. (1969) *Block or Day Release?* NFER.

'mathematics' (i.e. calculations) had causes and consequences which were important over a very broad spectrum of FE-related contexts certainly seemed to be of fundamental importance. The question was — would the idea stand up to systematic testing?

Ruth's job was to start on this testing, given very little money and hardly any time. She decided that the most important question was as to whether there was any firm evidence that students did have difficulty with some parts of their mathematics courses, and if they did, whether the nature of these difficulties could accurately be diagnosed. Having secured the enthusiastic collaboration of the City & Guilds of London Institute, she embarked on an analysis of the performance of nearly 18 000 students in CGLI examinations — and emerged from the ordeal with some pretty clear ideas about the kinds of questions they found difficult, and about the reasons for those difficulties.

That was the beginning. Since that time, giant industrial organisations such as BP Company plc and major research foundations such as Nuffield and Leverhulme, and a variety of governmental/quasi-governmental agencies, have supported the continuation and extension of the work. The problems of diagnosing and correcting difficulties with mathematics learning have been explored in schools from the primary level upwards, and among undergraduates, further-education students, and teachers-in-training.

In the furtherance of this very substantial undertaking, Ruth soon came to be working with Meg — or rather, with MEG, the Mathematics Education Group. The membership of this little group has changed gradually over the years, as has its size, but in one respect it has been invariant. Each member has combined a high level of competence as a mathematician, with an understanding of the statistical and methodological aspects of educational research. This is an extremely scarce combination of talents, and Brunel's department of education has been fortunate in being able to attract it so successfully. I am sure that the authors of this book would wish me to pay tribute to the invaluable contributions made by their colleagues over the years, and in particular to the indispensable help of Colin Winter, who has combined his membership of MEG with his role as the department's Mathematics Methods Tutor — to the great advantage of both.

Educational research is directed to the improvement of our ability to achieve educational aims, and in that sense it is often more concerned with applications and development than with theory. Ruth's early work, however, had a spin-off of a highly theoretical kind, which has substantially influenced our whole approach to the study of mathematics education. Instead of using multi-item tests as her main research tools, and groups as her main units for analysis, she has emphasized at every stage, the importance of studying the individual learner making the attempt to solve the individual problem. Looking at her data one day, with my psychologist's hat on, I suddenly realized that these were in exactly the form that was needed to make it possible to take a fresh look at a problem that had bothered psychologists for some time, but on which the evidence seemed to be equivocal. This was the problem as to whether mathematical ability was simply a particular expression of general intelligence, or whether it was a relatively specialized capacity in its own right. Working with her data, we came to the conclusion that although an adequate level of general intelligence plays a major role in enabling a student to learn and understand the individual operations of mathematics, it is not by itself sufficient to enable the student to bring these operations to bear in effective combination and sequence, on the solution of mathematical problems. Highly intelligent pupils quite frequently seem to lack this skill, even when the problems themselves (to the experienced mathematician) seem trivial.

The diagnosis and treatment of 'difficulties with mathematics' is thus a complex undertaking, and the expert mathematician may perhaps find it particularly demanding just because he *is* expert, and has perhaps forgotten how very counter-intuitive many mathematical operations seem when first encountered. Ruth Rees and George Barr have produced a volume which mathematics teachers at every level will find informative and of great practical value.

Prof. W.D. Furneaux
Director of Research
Department of Education
Brunel University

November 1983

INTRODUCTION

We have written this book to help you explore your students' understanding of maths. Finding out what goes in their minds is not easy and so we have tried to spotlight for you the kinds of tasks which really expose misunderstandings. We have also made some suggestions to help you cure these. Even experienced teachers need help with effective diagnosis because it is rather more subtle than it appears. Tasks which look the same and to the mathematician are the same, can present unanticipated difficulties to the learner, and not just to the slower learner but to the more able student as well. So this book is about diagnosis and prescription *across the ability range*.

We have worked with teachers and some 22 000 students from 10 to 57 years of age; talking to them, giving them written maths tasks and using the language laboratory to listen in to their thinking aloud as they solved these tasks. Starting in 1970 with students in further education it soon became obvious that we needed to extend the work to involve students across the educational spectrum in age as well as abilities: students at school, at university, trainees at work, the adult in everyday life. Is the misunderstanding of the 13-year-old that of the adult? Does what is learnt at school today matter tomorrow? So we looked at all these groups of people. What then did we find?

We found that there are certain maths tasks which cause difficulty to all students: the degree of difficulty, of course, depends on the age and ability of the student. We called this set of tasks 'the common core of difficulty'. On the basis that today's student may be tomorrow's teacher we carried out a study with trainee teachers. Did the same misunderstandings arise here? Is there a learning – teaching – learning cycle? Furthermore what of experienced teachers: are they aware of this 'core of difficulty'? (We now call it the 'core of difficulty' because the term 'common core' has come to be used in a different context.) And if so, what can be done about it? All this you will find discussed later in the book.

The core of difficulty considered relates to a limited set of mathematical topics involving certain fundamental concepts. These concepts though elementary are not trivial: they form the basis for more advanced mathematics. The Cockcroft Report proposes a foundation list of mathematical topics that all students should study but the core of difficulty will be buried even within this list. Diagnosis in these topics is essential if our teaching 'from the bottom up' is to be effective.

Asking the *right* question can unlock the blockage; *explicit* teaching may prevent the blockage arising. What teaching activities are effective for school students, college students and for adults? What is the effect of context? How do we take into account the effects of anxiety and attitude? These issues will be some of those discussed.

Undue concentration on learning difficulties can be counter-productive; effective diagnosis is *constructive*. Our increased awareness of the real nature of our students' misunderstandings can generate creative teaching on our part, increased competence, confidence and enjoyment on theirs.

HOW TO USE THIS BOOK

What you will find:

Chapter 1
This provides the background to the text and discusses why effective diagnosis is necessary for effective teaching.

Chapters 2 - 9
These chapters form the main body of the text. Each chapter deals with a specific 'core of difficulty' topic, and gives guidance for diagnosis and prescription. Guideline exercises are included that can be used with your students but which also serve as a basis for your own development of diagnostic tasks.

Chapter 10
The ability to solve problems is important but putting maths in a relevant context for our students is not necessarily a general panacea. This chapter explores the effect of both context and mathematical content and the interaction between the anxiety and the attitudes which are generated.

Chapter 11
The misunderstandings of the younger student at school are crystallized in the adult. The accompanying lack of confidence and anxiety require careful handling. This chapter discusses some of the approaches we have found helpful.

Chapter 12
Teaching maths really well involves some very profound thinking about what appear to be simple matters. We have to pick our way through conflicting demands, so the aim of the chapter is to improve our judgements not to provide ideal solutions.

In the first part of the chapter there are discussions of some issues related to the difficulties underlying the tasks. The exercises at the end indicate ways in which teachers may build on these discussions to improve their own diagnostic skills.

Diagnostic Assessment
This is a set of diagnostic tasks designed to cover all those concepts and skills discussed in the text.

Referencing
To maintain the continuity of thought in the discussion we have omitted the traditional way of referencing throughout the text. Any work to which reference is made is included in Suggestions for Reading on pp. 223-226. This section also includes some of the Brunel research reports and papers should you wish to read more of the work of the group.

Suggestions for using the book

- Read it through quickly.
- Look in depth at one chapter at a time.
- Work through the guideline exercises as suggested.
- Study the tables carefully: they contain a lot of information.
- Study carefully the Diagnostic Assessment on pp. 218-222.

   Work through the 'attractors'.

   Carry out your own research study and prepare for it as follows:
   Rate the degree of *exposure* you think your students have received for each task. (A three-point scale would do: 1, no exposure; 2, some exposure; 3, high exposure.)

   Predict the percentage success rate (and percentage popular wrong responses) on each task of your group of students or, in more depth, of each of your students. This study entails a lot of work but enables you to reflect on your own teaching of the tasks. The greater the effort, the greater the pay-off. It could of course be carried out as a group activity.

Work with the student

- Give the diagnostic assessment.
- *Look* at your students' written work on each task.
- *Listen* to the way they are actually solving the tasks.

- *Learn* How did your expectations match the reality?
    What are the misconcepts?
    Is there a mis-match between your teaching and their learning?
- *Prescribe* What can you do about the misconcepts?
    Use each chapter and its guideline exercises to help you diagnose, prescribe and explicitly teach to eradicate the specific difficulties you find.

You will observe as you progress through this text that, as with all important documents, it is essential to read the small print! We hope that you will find the book helpful in your teaching but in the end, as always, it is up to you. Our very best wishes go with you.

CONTENTS

The Mathematics Education Group   v
Acknowledgements   vii
Preface   viii
Introduction   xi
How to Use This Book   xiii

Chapter 1   The Need for Diagnosis   1

Diagnosis and the student   2
  Listening: the language laboratory and students' routes to solution   2
        mis-matches between teacher and student   4
  Looking: students' written work   5
  Learning: what do we learn from the looking and listening?   6
  Format and nature of the tasks set to students   6
  Core of difficulty and the subtlety of diagnosis   7
  Does what we learn today matter tomorrow?   9
Diagnosis and the teacher   11
  Trainee teachers: a learning-teaching-learning cycle?   11
  Experienced teachers: do we really know our learners?   12
  A mis-match between teachers' expectations and their students' performance   12
  Do effective diagnosis and explicit teaching help our students?   12
Diagnosis, learners and teachers   14
Summary   16

Chapter 2 The Natural Numbers	18
Diagnosis	19
Signs and symptoms: addition and subtraction	19
Looking and listening	20
Signs and symptoms: multiplication and zero	20
Looking at written work	20
Listening to routes to solution	22
Signs and symptoms: division and zero	23
Looking, listening: division	23
Looking, listening: division and zero	27
Learning: some general comments	29
The diagnosis	29
Prescription and aftercare	31
Suggestions	32
Some questions for discussion	33
Some quotations for discussion	34
Diagnostic exercises	34
Flexibility in thinking	34
Guideline exercise 1: multiplication of smaller whole numbers	37
Guideline exercise 2: multiplication of larger whole numbers	37
Guideline exercise 3: division by smaller whole numbers	38
Guideline exercise 4: division by larger whole numbers	39
Chapter 3 Small Numbers: Decimal Fractions	41
Diagnosis	42
Signs and symptoms: multiplication of decimal numbers	42
Looking: 'mixed' and small numbers	42
Listening: small numbers	43
Learning: small numbers	45
More looking, listening and learning	46
Signs and symptoms: division	48
Looking: 'mixed' and small numbers	48
Listening: small numbers	50
Learning: small numbers	50
The 'knock-on effect': squares and square roots of small numbers	51
Does treatment work?	51
The diagnosis	52

Prescription and aftercare	52
Suggestions and discussion	52
Some questions for discussion	55
Guideline exercise 1: multiplication of small numbers	56
Guideline exercise 2: division of small numbers	57
Guideline exercise 3: addition and subtraction of small numbers	58

Chapter 4   Small Numbers: Common Fractions   59

Diagnosis	60
Signs and symptoms: addition and subtraction	60
Looking: 'mixed' and small numbers	60
Listening: small numbers	62
Learning: some general comments	63
Exploring the understanding further: conversion from common into decimal fraction form	65
Looking, listening and learning: conversion	65
The diagnosis	67
Prescription and aftercare	67
Common fraction models and algorithms: addition and subtraction, multiplication and division	68
Some questions for discussion	71
Guideline exercise 1: conversion of common fractions into decimal fractions	71
Guideline exercise 2: addition and subtraction of common fractions	72
Guideline exercise 3: addition and subtraction in practical situations	72
Guideline exercise 4: multiplication and division of common fractions	73

Chapter 5   Percentage, Ratio and Proportion   74

Diagnosis: percentage	75
Signs and symptoms	75
Looking: a variety of tasks	75
Listening	78
Learning	80
The diagnosis	81

Prescription and aftercare	81
Suggestions and discussion	81
Diagnosis: ratio and proportion	83
Signs and symptoms	84
Looking and listening: ratio and proportion	84
Looking and listening: more proportion	85
Learning	87
The diagnosis	87
Prescription and aftercare	88
Suggestions and discussion	88
Ratio; ratio as a fraction; proportion	88
Some questions for discussion	91
Guideline exercise 1: percentage	92
Guideline exercise 2: percentage	93
Guideline exercise 3: ratio and proportion	93

Chapter 6   Shape and Form: Similarity   95

Diagnosis	97
Signs and symptoms: key concepts	97
Looking: squares, cubes, circles, spheres, area $\propto$ (dimension)^2, volume $\propto$ (dimension)^3	97
Listening: primary-school students	104
secondary-school students	106
undergraduate engineers	107
Learning	109
Further signs and symptoms: units of measure	111
Looking and listening	111
Prescription and aftercare	112
Teachers' comments	112
Suggestions and discussion	114
Concrete approach	114
Real-life applications	114
Appeal to imagination	115
Some questions for discussion	116
Guideline exercise 1: area concepts	117
Guideline exercise 2: volume concepts	117
Guideline exercise 3: units of measure	118

Chapter 7 The Circle and its Measurement — 119

Diagnosis — 120
  Signs and symptoms — 120
    Looking: area of squares and rectangles — 120
      volume of cubes and cuboids — 122
      circumference of circle — 124
      area of circle — 125
    Listening: π and area of circle — 128
    Learning — 130
  The diagnosis — 131

Prescription and aftercare — 131
  Suggestions and discussion — 132
    perimeter, π, practical activities, area of circle — 132

Some questions for discussion — 134
Guideline exercise 1: circle formulae — 135
Guideline exercise 2: calculating circle area and circumference — 135
Guideline exercise 3: perimeter, area and volume of straight-line shapes — 136

Chapter 8 An introduction to algebra — 139

Diagnosis — 140
  Signs and symptoms — 140
    Looking: 10-year-olds and equations — 141
      older students and equations — 142
      older students and substitution: $x^2 y$ — 143
      older students and $\frac{1}{x}$ — 145
      younger students and $\frac{1}{x}$ — 147
    Listening: solutions — 148
      $x^2 y ; \frac{1}{x}$ — 148
      comparison of $\frac{1}{x}$ tasks — 149
    Learning: routine versus diagnostic tasks — 154
    Student and teacher: is there a mis-match? — 156
    The diagnosis — 157

Prescription and aftercare — 157
  Suggestions and discussion — 157
    Generalization — 158
    Flexibility in solution — 158
    Practice — 159

Some questions for discussion	160
Guideline exercise 1: substitution	161
Guideline exercise 2: equations with $\square$ and x	162
Guideline exercise 3: equations with $\frac{1}{\square}$ and $\frac{1}{x}$	163
Guideline exercise 4: forming equations	164

Chapter 9   Statistics: On Average 166

Diagnosis	167
Signs and symptoms	167
Looking: average; mean, median and mode	167
context	168
frequency table	169
Listening	170
Learning	170
The diagnosis	171
Prescription and aftercare	171
Suggestions and discussion	171
Some questions for discussion	172
Guideline exercise 1: mean, median, and mode of data not given a frequency table	173
Guideline exercise 2: mean, median, and mode of data given a frequency table	174

Chapter 10   Problem Solving: Context, Attitude and Anxiety 175

Context	176
Looking: one task in different contexts	176
mechanical context and mechanical maths	179
mechanical context, more demanding maths	180
inferential context, what of the maths?	181
inferential context, inferential maths	182
Attitude	183
The usefulness of maths	184
Liking and performance	185
Dislike and anxiety	186

Anxiety, attitude and performance	186
The study	187
Looking: mathematical solutions	187
Listening: mathematical solutions	187
Learning: attitude, anxiety and performance	189
Prescription and aftercare	190
Suggestions and discussion	190
Some questions for discussion	190

Chapter 11   Adults and Numeracy   193

Diagnosis	193
The first few meetings	196
Looking: the learning environment	196
Listening: to the students' feelings about maths	196
Learning	196
Diagnosis: informal and formal assessment	196
Prescription and aftercare	198
Activities after the first few meetings	198
Listening: ideas for discussions	198
Looking and listening: simulation exercises	198
Learning	198
Lesson plans: the structure and aims of a session	199
Recording and evaluating progress: a student and tutor activity	199
Some questions for discussion	200

Chapter 12   Teaching Maths   201

Training for Teaching	201
Different needs of preservice and inservice teachers. Implications of studying diagnosis for curriculum planning and for effect on teaching styles.	201
Diagnostic Questions and Diagnostic Tests	202
Distinction between diagnosis for better teaching and diagnosis for student grading and placement.	202
Writing Questions	204
The multiple-choice format and its justification for the form of diagnosis discussed.	204

Images and Consequences   206
   Mathematical abilities. How the route to a solution depends on the initial image or model conjured up by a question.   206

Routines and Imagination   208
   Relations between imaginative and algorithmic solutions. Revising our view of the place of the algorithm.   208

Context   208
   Does relevance and real-life context provide the answer to our difficulties?   208

Activities for Teacher Training   210
1. Analysis and comparison of individuals' methods and solutions, images and techniques.   211
2. Justifying the alternatives in multiple-choice questions.   211
3. Relations between computational skills and mathematical concepts.   212
4. Fraction models and fraction algorithms.   212
5. Examining and commenting on books and their variations in use and definition of fraction, percentage, ratio and proportion.   214
6. Dropping the level of abstraction of some common concepts.   216
7. Multiple-choice item construction.   216

References   217

A Final Comment   218

Diagnostic Assessment   220

Suggestions for Reading   225

CHAPTER 1

THE NEED FOR DIAGNOSIS

'Diagnosis... The identification of a disease from examining its symptoms.'

(Oxford Dictionary)

Most of us want our students to feel that maths is an enjoyable and rewarding study for them... but do we succeed? Some of our students are successful whereas others confidently and cheerfully keep getting things wrong; yet others are anxious and fearful. They can fool us by their apparent ability to solve many tasks correctly. Yet if or when the understanding is probed we realize that they have not the foggiest understanding of *why* they are doing *what* they are doing. So what goes wrong?

If we meet them later at school, college or on an adult-retraining course, sometimes we despair at how far back we need to go if we are to help them understand and make further progress. Thus, just as in health care, early awareness of 'blockages' is important for learners and teachers.

In medicine, despite the expertise of an experienced doctor, diagnosis is complicated by the fact that several diseases may exhibit common symptoms. Similarly diagnosis of mathematical misconcepts (we use 'misconcept' to mean inadequate concept or blockage) may also be a more subtle art for even experienced teachers than appears at first sight. Consider, for example, one of the core of difficulty topics, 'operations on numbers less than one'. Will *all* such tasks be diagnostic? Will they probe the understanding and locate the blockage?

Effective diagnosis in this topic reveals the formation of misconcepts in:
  Place value: the use of digits in the decimal system
  The meaning of zero as a number and place holder
  The meaning of multiplication.

The misconcepts diagnosed from exploring the understanding with certain critical tasks may thus stretch back in time to earlier learning and if not checked will spread, like a disease, in future learning. Early diagnosis as in the medical analogy is thus highly desirable not only for remedial treatment but preventative.

Diagnosis and the student

Listening

Monitoring routes to solution by talking aloud.
'Sometimes he thought sadly to himself, "Why," and sometimes he thought "Wherefore," and sometimes he thought "In as much as which," — and sometimes he didn't know what he *was* thinking about.'
(A.A. Milne *Winnie-the-Pooh*)

It became clear during the early years of studies at Brunel that *finding out what went on in a learner's mind* was crucial to effective diagnosis. We therefore decided to record students' 'routes to solution' in two ways. First, there is the 'interviewer-free' situation in which the student talks through the solution without interference. Secondly, there is the interview situation in which there is a dialogue between interviewer and student.

The aims of these two methods can be quite different and if one is seeking to find out students' routes to solution unhindered by interference from the interviewer then clearly the first method has advantages. For the purposes of research we selected this as the more effective initial diagnostic method since it enabled misconcepts to emerge without interference. The language laboratory was found to be ideal for this and it also had the distinct advantage

of simultaneously monitoring several students. If a language laboratory was not available or where school students had not been accustomed to its use then we attempted to reduce any interference to a minimum by asking them to talk through their solutions into a cassette-recorder. When we judged it appropriate and wished to probe further the interview situation was used.

However students' recordings were carried out, whether in the language laboratory or speaking into a cassette-recorder, they were asked to speak out loud any thoughts at all that came into their minds; to give a detailed commentary just as if they were describing the route taken when riding a bike. If they got confused or lost or went blank or didn't know what to do or felt anxious or panicked we asked them to recall all their thoughts and feelings on the way to their solution of a task. The language laboratory was particularly effective and quite a lot was revealed from the detailed commentaries! Isolated in their booths students became uninhibited, revealing all their frustration, anxiety, confusion and, even, confidence.

There were points of discontinuity where students became lost. Sometimes they went back to the beginning and checked their solution again. It was helpful to see which students checked their solutions or were not happy with the look of the solution just as a musician may be unhappy about the sound of a wrong note.

The mis-match

Several aspects of solving tasks which are not revealed by the written situation have been highlighted by the work in the language laboratory. One of these is the 'mis-match' which can appear between teacher and learner. One kind of mis-match is that in which a learner sees a task quite differently from that of the teacher or different from the way in which the teacher assumes the learner is thinking.

Another kind is the mis-match discussed by Richard Skemp which can arise if learning is instrumental (rules without reasons) and teaching is relational (understanding not only *what* but *why*) or vice versa. We have all no doubt been frustrated at times when we have been trying to teach relationally only to discover that our students want to learn instrumentally; they want a quick answer.

Task If $\frac{1}{x} = \frac{3}{4}$, x is?

How the learner sees the task   How the teacher sees the task

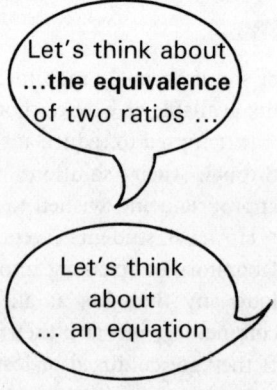

Craft student — *x on the bottom... ??? Help!!!*

School student / Technical student — *x? What works?*   *Cross-multiply*

Teacher — *Let's think about ...the equivalence of two ratios ...*

Let's think about ... an equation

Instrumental learning?   Relational teaching?

'*Impact perception*' is the way in which the learner first sees the task and is another seemingly important aspect. *This can colour the whole route to solution.* Getting off to a good start is important! We are only too familiar with the tenacious rigidity with which first impressions hold; names wrongly associated with faces on introduction can cause continued embarrassment much later! And the analogy holds with all learning; perhaps mathematics with its concise symbolism is particularly vulnerable. This is why explanations, which we use to help our students acquire a concept, need very careful consideration.

One example, that we cannot resist quoting, is taken from a well-established curriculum work card on introducing fractions.

'Silly Basil.
Basil Brayne said that $\frac{1}{4} + \frac{1}{2} = \frac{2}{6}$
because $1 + 1 = 2$ and $4 + 2 = 6$.
Explain carefully why Basil was wrong and give the correct answer.'

It may be that despite the instruction the superficial *impact perception* here may haunt a student's future thinking and working.

Looking: written work

In diagnosis one important aspect is the development of appropriate diagnostic tasks. Such tasks have to be both related to the curriculum of the learners being studied and effective for diagnosis. It is not easy to meet both these requirements. When we began our work most validated written tests that were available measured success; that is they settled for the first requirement and rejected those tasks on which learners performed badly. Yet we felt strongly that those very tasks on which we found learners performing badly may well be those which probe the understanding and expose the misconcepts; this initial conviction has been more than justified.

It is also essential that the written questions should be clear and unambiguous if we are to make proper use of the information obtained from testing. Both teachers and researchers can fall into the trap of setting tasks which require learners to 'guess what is in our minds'! Yet as the Cockcroft Report points out: *'Mathematics provides a means of communication which is powerful, concise and unambiguous...'*. We write quite strongly on this point since the language-laboratory recordings have shown the considerable mismatch which can occur between how the student sees a task and what the teacher/or researcher thinks has been taught or thinks has been asked.

We deliberately selected the multiple-choice format as a diagnostic instrument for the written tasks. (This format asks the student to choose between four or five possible answers of which one is correct and the others are called 'distractors'.) The distractors (we prefer to call them 'attractors') were carefully designed to reflect common ways of incorrect thinking which were based on substantial analysis of students' performance together with our colleagues' and our own combined experience of teaching.

Learning

The technique that we pioneered and have used since the early 1970s of combining the complementary approaches of the written and talking-aloud procedures has proved a powerful diagnostic method. The method can be reproduced by teachers in the classroom using written diagnostic material and then listening and talking to their students: in other words, putting into play the motto for both learner and teacher 'look, listen and learn'.

If a language laboratory is available then its occasional use for maths diagnosis could act as a stimulant to teaching and learning. The student is in a one-to-one situation with the teacher without fear of classmates eavesdropping and both teacher and student have *immediate* feedback. All too often the effectiveness of correcting written work is diminished by the time which can occur between our marking the work and returning it.

For day-to-day teaching in the maths classroom the interview situation is probably more practicable and could follow fruitfully from a 'one-off' language-laboratory period. The teachers' role in activating discussion about mathematical thinking is crucial; discussion of tasks enables students' thinking to 'loosen up' and helps them to see that the mathematical formulation with symbols is a concise, efficient and powerful working language... and is not black magic!

What else did we learn?

Format and nature of the tasks

The language-laboratory recordings revealed that some learners 'did not like the answers provided'. We therefore modified the multiple-choice format of the tasks and included a space for the student's own response having made it clear that there was not necessarily a right answer among the distractors provided. Appropriate design of such tasks enables 'the best of both worlds' to be obtained: the learner works through the task as if it were 'open-ended' and the teacher can have the advantage of objective marking.

The written tasks were carefully designed and based on considerable analysis and discussion. We had frequent visits with teachers in schools and lecturers in colleges to consider the mathematics they were expecting their students to know and to be able to apply. We had also analysed the performance of many thousands of further-education students on a national examination.

The earlier tasks were concerned with the basic concepts and skills associated with number and the beginnings of algebra and geometry; the kind of mathematics that is essential for application in the real world or any further mathematical study.

These sets of tasks were given across the educational spectrum from 10+ to adults of 57 years and it is the results of this work which form the basis of this book. We have since developed further sets of tasks with more advanced mathematics which we have given to smaller groups of students and the results consistently confirm how essential it is to lay a good foundation with basic concepts and skills.

Core of difficulty and the subtlety of diagnosis

Three of the findings relating to learners are:
● Certain misconcepts are shared by mathematics learners of all ages.

These common misconcepts are labelled the 'core of difficulty' and are associated with the following topics:

The natural numbers and zero : multiplication and division
Small numbers (between 0 and 1) : multiplication and division
The 'inverse'
Measurement of area and perimeter : the circle
Similarity : properties of similar figures
Context in problem solving

For the purpose of this book, these misconcepts will be limited to this particular range of topics as pointed out in the Introduction. Nevertheless, our later work, using tasks designed with more advanced mathematics, has confirmed how essential it is that these elementary yet fundamental concepts are firmly established. Their effects spiral out into many more complex tasks.

The difficulties associated with some of these topics are now being explored by other workers both in this country and abroad and it is being confirmed that these misconcepts are fundamental in that they occur with students irrespective of the educational system or curriculum. There is now emerging substantial information on difficulties encountered in learning and this is where effective diagnosis can be of considerable help to the teacher.

- Diagnosis of difficulties is subtle.

Changing the context of presentation of a mathematical task in ways which seem superficially to be quite small may very substantially change the nature and severity of the difficulties a student experiences in attempting to complete it.

Examples of the subtlety of diagnosis are given below. Look carefully at the two types of tasks: they look so similar and yet to the learner they present quite different kinds of difficulties.

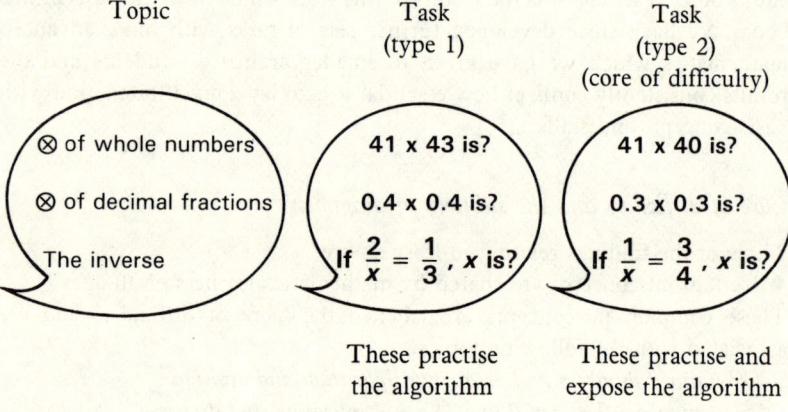

These practise the algorithm

These practise and expose the algorithm

Algorithm: process or rule for calculation.

Do you see why students are apparently more successful with tasks of type 1 than they are with tasks of type 2? It is not sufficient to know in which *topics* the difficulties are buried: the subtlety lies in knowing which *tasks* will expose a learner's 'rules'. Tasks of the first type are amenable to rote application of a rule. Tasks of the second type (the core of difficulty) are also amenable to a routine but this routine must be used with understanding.

- Students, even with an apparently sound grasp of a range of mathematical operations, are sometimes quite unable to use and integrate those operations within practical and/or theoretical problem-solving activities. As an example of this consider the following task:

Task $8 \div 0.16$ **is?**

Given that resistance $= \dfrac{\text{potential difference}}{\text{current}}$
an appliance carrying a current of 0.16 amperes with a potential difference of 8 volts has a resistance of?

The first task was in a 'pure' computational form; the second involved the same mathematical operations but in a context *relevant* to the particular learner, i.e. craft and technician students. There are difficulties associated with the first task because it involves division by a decimal fraction (a 'core of difficulty' task). Nevertheless even students who would tackle this task using a well-established routine are not necessarily able to tackle the second successfully because it involves knowing *what* concepts, *why* they apply and *when* to use them.

Does what we learn today matter tomorrow?

Every teacher knows from how far back an older student's difficulties can stem. There is now overwhelming evidence to enable us to state with conviction that the misconcepts formed by the 13-year-old can indeed be those of the adult. Some examples of this for core of difficulty tasks are given in Table 1.

The misconcepts of the adults on retraining courses can be accompanied by feelings of inadequacy and lack of confidence which can be traced back to school. We found that, when they were faced with tasks similar to those given at school, strong memories were aroused:

'... most of the trainees... had an illogical fear of sitting a written test.... Most of those who coped better in the discussion revealed that they had been poor at maths and unconfident at school where the emphasis is often placed on absorbing enough in a short time to achieve an acceptable standard in exams. The pressure and subsequent failure of some adults resulted in a strong conviction of their own inadequacy....'.
The common core tasks form an essential foundation of what follows later. At one extreme they are needed for the minutiae of daily living and for an appreciation of the world around us. At the other extreme they are essential for students' continued mathematical development as we shall be discussing in later chapters.

10 Diagnosis and prescription in the classroom

Table 1 Teachers' predictions of their students' success rate compared with the reality

Tasks	Teachers' estimated % success rate for 13/14	Actual % success rate for 13/14		Actual % success rate for TOPS (19–57 years)	Popular wrong response common to 13/14 and TOPS
23)4669	64	46	**38**	39	23
513 ÷ 27 is?	69	71	**39**	73	17
0.3 × 0.3 is?	51	45	**33**	37	0.9
0.7 ÷ 0.7 is?	54	29	**6**	27	0.1
$\frac{7}{16}$ correct to two decimal places is?	41	13	**8**	21	7.16 & 0.43

Numbers in **bold type** refer to slower learners
Students are 13/14-year-olds and 'TOPS' trainees.

Diagnosis and the teacher

Trainee teachers: a learning–teaching–learning cycle?

'Today's student is tomorrow's teacher': it was on this basis that we extended our work to include 600 teachers in training. We had set out to explore the understanding of students at primary and secondary school, further education college and university ... but what about the student teachers who tomorrow would be in the schools?

The student teachers were mostly non-specialist mathematicians training for the primary and secondary sectors and the majority were at the end of their second year of a 3-year course. They had recently taken a mathematics examination and could therefore be said to be at the peak of their performance. Also taking part in the study was a group of further-education lecturers on an inservice training course.

Looking

The teacher trainees were given the same tasks as all the other students; once again the 'common core of difficulty' did its exploration convincingly. These tasks were found to be difficult *and the most popular wrong responses were the same as those selected by the other learners*. For example, only 22% of the primary-school trainees could select the correct answer to 'the square root of 0.9 is?' with 42% thinking that it was 0.3; only 17% could select the correct answer for the area of a circle.

The misconcepts were shared, although not necessarily to such an extent, by secondary school trainees and some inservice business-studies lecturers.

Listening

'I have never been able to do mathematics. I dislike mathematics'.
'I like mathematics but cannot do it'.
'I now tolerate mathematics but hated it at school'.
'I have done quite a lot of mathematics at college but I am scared of teaching number to young children'.

The results of the study and the comments of the trainees may seem to present an unduly depressing picture but nevertheless it reflects reality, judged by the many follow-up studies or students' projects which this particular one stimulated around the country. The optimistic view is that once there is effective diagnosis there can be effective prescription.

Experienced teachers: do we really know our learners?

In his discussion on communication in teaching and the evoking of concepts by symbols, Richard Skemp says 'We may think we are communicating when we are not... .'

We had the opportunity to carry out a study with teachers in which their expectations were linked to their students' performance on the diagnostic material provided. These students were 13/14 years old and spanned the ability range. Figure 1.1 shows the expectations compared with the reality. Figure 1.2 is an alternative way of showing this information. It will be seen from the figures that students' success rate for many of the core of difficulty tasks is significantly lower than that predicted by their teachers. The subtlety of diagnosis is spotlighted by the two tasks selected which look so similar and yet produce such different reactions.

These results were very similar to those obtained and reported some ten years previously with lecturers in further education and their craft and technician students.

There are other tasks for which both school teachers and further-education lecturers *overestimated the difficulties*. This too is useful information, for all of us find that time is one of the most important resources in planning our teaching programme, and we may not wish to over-teach.

Do effective diagnosis and explicit teaching help our students?

Our teaching colleagues felt overall that they were made more *aware* of the need

To ask pupils the right questions to probe understanding
To appreciate the nature of their pupils' difficulties
For explicit teaching of the relevant concepts and skills.

The general feeling was that, although the diagnostic materials had been used in the time available to concentrate on certain basics, nevertheless the involvement had reorientated their whole approach to the teaching of mathematics. By being made *aware* of some of the misconcepts which can arise with third-year students they reassessed the implications for the 'spiralling' process that goes on in later years.

In the limited time available our teaching colleagues carried out their own chosen programme of *explicit* teaching of the concepts and skills associated with the core of difficulty tasks. The results were encouraging: students' overall performance on the tasks improved. Perhaps the most important effect

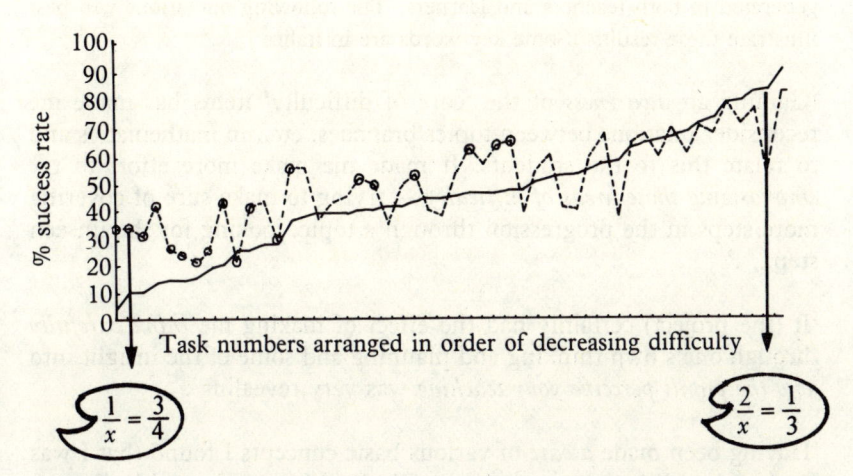

Figure 1.1 A mis-match between teachers' expectations and their students' performance. (———, Actual; ----, predicted; O, core of difficulty task.)

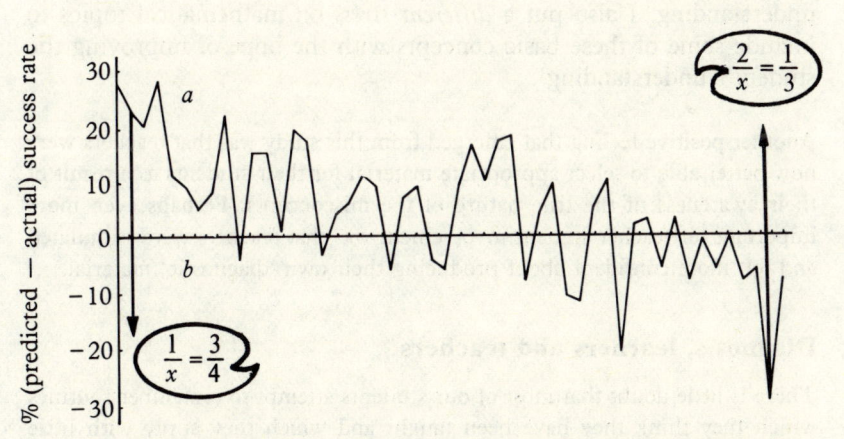

Figure 1.2 An alternative way of seeing the difference between predicted and actual success rates. (*a*, Difficulty underestimated; *b*, difficulty overestimated.)

was not only the *improved competence* of students but the *increased confidence* generated in both teachers and learners. The following quotations will best illustrate these results. (Some key words are in italics.)

'Gaining an *awareness* of the 'core of difficulty' items has made me reconsider relations between topics/branches, etc., in mathematics and to relate this to the students. It made me make more effort to try *emphasizing those areas of difficulty* — trying to make sure of covering more steps in the progression through a topic, looking for the unseen step... .'.

'It (the project) certainly had the effect of making me *think carefully* through one's own thinking and planning and some of the insight into *how the pupils perceive your teaching* was very revealing'.

'Having been made *aware* of various basic concepts I found that I was *teaching in a different way*, not just with the third year but with all years including a fourth-year 'O' level set. I became aware that I was assuming too much previous learning and understanding and I feel I now have a better idea of *when and how* to *question* a student's lack of understanding. I also put a *different stress* on mathematical topics to include some of these basic concepts with the hope of improving the student's understanding'.

Another positive feeling that emerged from this study was that teachers were now better able to select appropriate material for their students as a result of their awareness of the true nature of the misconcepts. Perhaps even more important for teacher-student involvement was that teachers were stimulated and felt more confident about producing their own 'diagnostic' material.

Diagnosis, learners and teachers

There is little doubt that most of our students attempt to remember routines which they think they have been taught and which they apply with little understanding. There is not much evidence of mathematical thinking (looking for relations, patterns, generalizations) and this applies not only to school and college students but to university undergraduates. Mathematics for most students is reduced to being essentially algorithmic in nature. And yet those of us who have taught younger children know how creative their thinking can be; those of us who have probed deeply in work with craft students, some of whom were 'failures' at school, have found evidence of genuine mathematical thinking. Is the algorithmic formula a reflection of our own teaching... of the way we were taught? Is it our response to young people's impatience of wanting to get things done quickly... is it easier? Do we pressurize ourselves with 'getting through the syllabus' as our prime aim... is this a defence against pausing and reflecting, in our hectic lives?

Effective early diagnosis can be a great saver of our time; it can avoid many of the later pitfalls which could occur in learning. The *prescription* which should follow involves turning the microscope on ourselves: thinking more carefully of the tasks we set, looking and listening to the way they are carried out, watching the language we use to explain mathematical ideas. We do not have to lose mathematical rigour if we use the appropriate words; the short term pay-off which may occur if we are tempted to sacrifice the true language of maths does not compensate for the later confusion it may cause.

The teaching–learning interaction is complex: there are no easy answers and there are important bridges to build. Just how do we teach? We cannot teach a concept. Z.P. Dienes recommends that we help our students abstract the concept from a variety of examples both mathematical and perceptual. On the other hand, if we teach for algorithmic skill the tendency is to stick to one specific method; do we confuse the student if we don't? The best solution is surely to build a bridge between the two approaches so that our students both acquire the concept and the skill.

We would plead for *flexibility* in teaching (there are several ways to solve a problem) thus opening up the understanding; combined with sufficient practice *well-understood routines* can be established and give confidence. We would wish this for *all* learners, even the slower ones, so that all are exposed to the power, the generalization, and the excitement of mathematics. It's not easy but it is worthwhile.

Both learners and teachers need to	LOOK LISTEN and LEARN
As teachers we need to be	AWARE EXPLICIT and FLEXIBLE
so that our students may have the	CONCEPT COMPETENCE and CONFIDENCE

Aspects of Secondary Education in England (HMI report 1979) possibly sums it up for us by saying:

'Practice needs to be carefully controlled and supplemented by

i improved diagnosis of pupils' individual difficulties
ii a better appreciation of the role of language and oral work
iii more effective use of the applications of the ideas, both in the world around and in other subjects in the school'.

Effective diagnosis...   *Effective prescription...*   *Effective learning?*

Task   0.3 x 0.3 is?

Learners   Teachers

0.09   0.09   0.09   0.09

Yes!   Understood?   Yes!

Summary

Effective diagnosis is
- Necessary for effective teaching.

Our students can fool us by their apparent ability to carry out routines successfully in many cases

- Best carried out with both the written and 'talking-aloud' procedures: 'Look, Listen and Learn'.

- Subtle.

Tasks may look similar but certain will reveal the blockage, others will not.

The first kind, the algorithmic tasks, are amenable to 'rote application' of an algorithm: they demonstrate the kind of understanding called by Richard Skemp 'instrumental understanding'. The second kind, diagnostic tasks, are also amenable to the algorithmic solution but they expose the understanding of the algorithm used, rather more: i.e. more 'inference' or 'deduction' is required for solution. They may well demonstrate Skemp's 'relational understanding'.

This chapter may perhaps best be summed up in terms of the characteristics of these two kinds of tasks.

Algorithmic tasks

Students perform reasonably well provided they recognize the problem

Teachers' estimates of students' performance are reasonably accurate

Diagnostic tasks

Performance is relatively worse for all kinds of students

Even experienced teachers often do not appreciate the nature of the difficulties and underestimate the extent of the difficulties
Non-specialist teacher trainees often have the same misconcepts as other students: indicating a learning–teaching–learning cycle?

Effective prescription has to be directed towards treating:
  An overall lack of feeling for the symbolism and language of mathematics
  Specific deficiencies
  Mis-matches between learner and teacher.

This can best be effected by teachers being   AWARE EXPLICIT and FLEXIBLE
to develop learners'   CONCEPTS COMPETENCE and CONFIDENCE.

CHAPTER 2

THE NATURAL NUMBERS

'God created the natural numbers... .'

(Kronecker)

In general, people feel more at ease with the natural numbers, the counting numbers, than with any other form of mathematics. Yet we still require a formal system for writing these numbers and working with them. And understanding this system is where learning difficulties occur. It is perhaps

"Look—if you have five pocket calculators and I take two away, how many have you got left?"

Punch, October 13 1976

not surprising in view of the fact that the decimal system, where we count in tens, took man several hundred years to develop.

Helping our students 'acquire a feeling for number' must surely be one of the best ways to lay a sound foundation for their future mathematical learning and achievement. To do this we ourselves need to be aware of the learning pitfalls which could occur. It will clearly not be possible to discuss all the difficulties which could arise with our students across the ability range. Slower learners, for example, whether at school, college or on retraining courses will tend to have 'more of the same' of the core difficulties.

In this chapter, therefore, the emphasis will be on *spotlighting* those tasks in *multiplication* and *division*, especially those involving the use of *zero*, which we have found useful in exploring the understanding of most learners. Such tasks will in turn harbour the misconcepts with addition and subtraction.

Our students young and old can sometimes fool not only themselves but us: their apparently successful use of an established routine (algorithm) reflected in a set of correct answers can lull us into a false confidence of their understanding. The diagnostic probe is needed to reveal their misconcepts and guide our teaching.

Diagnosis

Signs and symptoms: addition and subtraction

Addition and subtraction of the natural numbers in a vertical format with *no* carrying is a routine which, as we are aware, most of our students carry out successfully. Carrying, however, explores the understanding rather more.

Tasks may look alike but...

$$+\frac{135}{241} \qquad -\frac{387}{124} \qquad +\frac{135^*}{246} \qquad -\frac{387^*}{129} \qquad \text{*Diagnosers}$$

These practise an algorithm ··· These practise and expose
                                              the algorithm

Diagnosis and prescription in the classroom

Looking and listening

Do you find these rules being practised?

'Always take the smaller from the larger'

$$\begin{array}{r} -387 \\ 129 \\ \hline 262 \end{array}$$

Signs and symptoms: multiplication and zero

Looking
Is this what you see?

Task 7 x 8

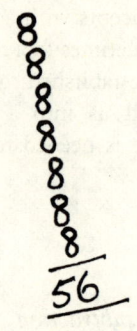

$$\begin{array}{r} 8 \\ 8 \\ 8 \\ 8 \\ 8 \\ 8 \\ 8 \\ \hline 56 \end{array}$$

We found this method being used by 17+ craft trainees. It shows basic understanding; you should use it.

Task 403 × 12

$$\begin{array}{r} 403 \\ \times 12 \\ \hline 403806 \end{array} \qquad \begin{array}{r} 403 \\ \times 12 \\ \hline 806 \\ 4030 \\ \hline 12090 \end{array}$$

Getting lost in the mechanics of long multiplication: one rule is as good as another!

Look at this one; this is what the examiners saw from 52% of 8613 candidates in a recent public examination.

$$9 \times 0 \times 7 = 72$$

Task   17 × 100

```
   100           17            17
    17         ×100          ×100
   100          170           700
  1000           70
  2000           00
                170
```

Drowning in a sea of zeros; long multiplication rules! (OK?) Let's hope these students don't become radiographers/nurses/buyers/electricians.

How then can we best diagnose misconcepts arising with place value?

Two similar multiplication tasks like these can be revealing:

This triggers an algorithm   *This* raises doubts about the algorithm

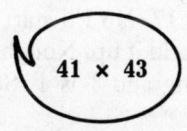

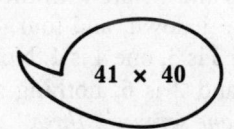

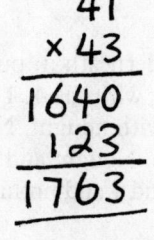

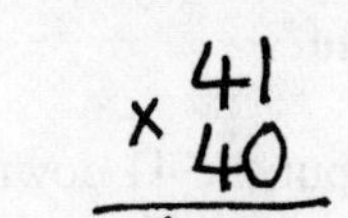

(41 × 43)   (41 × 40)

Long multiplication seems   but ...   often breaks down here (do I
OK here?   really understand?)

```
    41                    41
  × 43                  × 40
  1640                  1640       Multiplication
   123                    41       by zero
  1763                  1681       not
                                   understood
```

22 Diagnosis and prescription in the classroom

Task 41 × 40

```
   41                              41
 × 40         Role of            × 40        Mechanics
 ────         zero as            ────        in action!
  164         a place            1640
 0000         holder             0000
 ────                            ────
  164                            1640
```

We gave this one to our 10-year-olds; half of them got it right, the other half produced 63 different answers.

Both tasks can be successfully performed with long multiplication but it is not obvious that this indicates understanding. This is where listening is so helpful.

Listening
Take the case of 13-year-old Mechanical Micaela:

Task 41 × 43

'I put the 43 down and I put the 41 down underneath, put a times sign, put a cross and I start with the 4, so I put a nought down. Four 3s are 12, put the 1 down, and four 4s are 16 plus 1 is 17. No I'll start with the 1. One 3 is 3, one 4 is 4. Now I put a line and add it up. Nought and 3 is 3, 4 and 2 is 6, nothing and 7 is 7. Nothing and 1 is 1. So the answer is *one seven six three.*'

and

Task 41 × 40

'First of all I put the 41 down and put the 40 underneath it, put a line and I go put a nought down because we're starting with the 4. 4 times 1 is 4, 4 times 4 is 16, put another line and I start with nought. Noughts 1 is nought, noughts 4 is nought. Now I add it up. Nought and nought is nought. Nought and 4 is 4. 6 and nothing is 6 and 1 and nothing is 1. So the answer is *one six four nought.*'

Micaela has got the mechanics right but expresses the answers as separate digits. Can we learn anything from this?

Too few students show an understanding of zero; 12-year-old Diagnostic Deborah may be one of the few. Having worked out 41 × 43 by long multiplication she realized that there was a quicker method for the second task.

Task 41 × 40

'Well, I might times 41 by 4 and then times it by nought... it's quicker. Because it's a 10, a multiple of 10 and so I could times by 4 and then times by 10.'

This sounds hopeful, but does she really understand?

Signs and symptoms: division and zero

Looking
Is this what you see?

Task 28 ÷ 4

28 divided by 4
                               11-year-old

4, 8, 12, 16, 20, 24, 28,
            and that's 7

                         2013
                      3⟌639   10-year-old
                         6
                        ───
                        039

... use this understanding.

The natural numbers 23

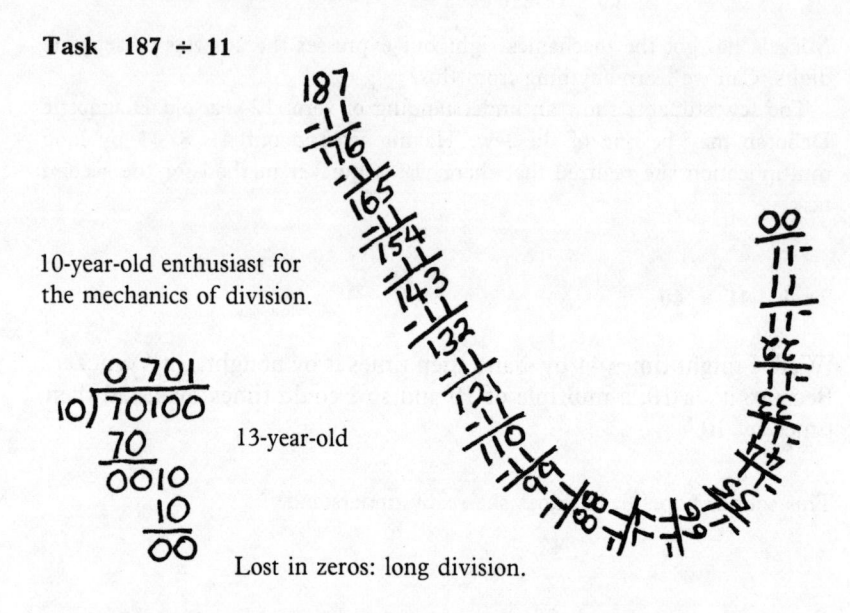

24 Diagnosis and prescription in the classroom

Task 187 ÷ 11

10-year-old enthusiast for the mechanics of division.

```
   0 701
10)70100
   70
   0010
     10
     00
```

13-year-old

Lost in zeros: long division.

We wanted to find out how flexible students could be in their thinking. Would they always apply the algorithm of long division mechanically or would they look for quicker methods? We found that 513 ÷ 27 is a good diagnoser of a 'feeling for number'.

Task 513 ÷ 27

Long division appeared to rule!

The 10-year-old students gave 105 different answers, the 12-year-olds 45... things get better! As the age of the student increases the long division algorithm appears to be used with greater effect as seen in Table 2.1... but is it understood?

Table 2.1
Task 513 ÷ 27

Students	% getting correct answer
	19
Primary 4th-year junior	32
Secondary 2nd year (mixed ability)	43
Secondary 3rd year (mixed ability)	71 **39**
Secondary 4/5th year (mainly 'O')	70
Craft/technician 17+	
Precourse	63
Postcourse	73
TOPS trainees (19–57 years)	73

Numbers in **bold type** refer to slower learners

Some students do not like showing their working on paper. 'I prefer to do it in my head' was said not infrequently. And so it was the 'thinking aloud' of the students in the language laboratory which was more revealing than their written responses.

Listening
The language laboratory and interview recordings.
For many students, memory of the mechanics of long division had dimmed:

Task 513 ÷ 27

13-year-old schoolboy:

'..., 27 into 13... put three noughts after 513 to take some noughts down, 27 into 51... no, don't think it does ...no... so that's um 51... 11 into 7 is 4, 1, 27, won't go put up 1, 11 × 6? ... means (working on paper) ... long pause... um. I think... answer is 11. NOT sure because its a bit hard.' (NB Long time spent on this.)

26 Diagnosis and prescription in the classroom

13-year-old schoolboy:

'...oh, how do you do it by long division, can't quite remember. Do it separately, 7 into 513 equals, oh! 7s into 51 goes 7 remainder 2... (works it out)... (long pauses).... Can't work it at moment, got to think about it. Uh...um... 27 times 10 equals 270, 270 times 2 equals 540, yeah, 27 times 20 is... oh... 23 point something' (long pause) (blurred comment).

This student could not remember the method of long division so decided at first to divide 7 and 2 separately. Was he using the model of long multiplication as an analogy?

13-year-old schoolboy:

'513 divided by 27 is 27 into 51 goes... 1... with 24 over so that's 27 into 243... goes...er... oh... goes 9. So the answer is 19.'

... Success!

First-year university engineering student:

'513 divided by 27 — well 2 x 27 equals 54, goes once, leaves 13... 27 into 240 something like... 10 times gives 270. Take 54 off 270... what is it... 18.8 remaining... haven't got this answer down so reckon I'd better check it... by multiplication... so must go 18.8 recur so 18.89 to two decimal places.'

... Nearly there... but the bridge fell down!

First-year university engineering student:

'Let's see if anything divides in both... 3 or 6... 3. 513 divided by 3...oh...etc. 57/3... oh... 3 × 9s is 27... eh... answer 19. No question about that!'

...Cheers! A quick method: this student was one of those who could be counted on one hand.

Interviews (I, interviewer; S, student; for this and proceeding chapters)

Task 513 ÷ 27

17+ craft student:

I: Any idea?
S: Divide... um... .
I: 513 divided by 27.
S: 27 into 513.
I: In your head?
S: Um. Probably start with 1 and work all way up.
I: Yes! You mean 27 by 1, then by 2... all in your head.
S: Take a long time. Dunno how to do it on paper.

This student liked working things out in his head. He had not taken kindly to working out sums on paper even when at school.

17+ craft student:

S: 27 into 513... .
I: How will you work this one out?
S: 27 into 5, won't go, 51 go once, 7 from 10 is 3 and 1 is 4, 243, 27 into 24— won't go — bring down 3, 27 ...243... long pause.
I: Trying 27 by 17 to see if it fits?
S: Yes (multiplies 27 by 17 and gets 459)... .

This student shows an understanding of multiplication and division as inverse processes, a good basis on which to build.

How then can we best diagnose misconcepts arising with division and also involving zero?

Here are some further suggestions.

Task 842 ÷ 2   **Task 812 ÷ 2**

These *look* alike:

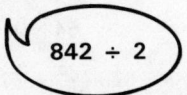

$\quad$ 842 ÷ 2 $\qquad\qquad$ 812 ÷ 2

The routine used here   ...   may break down here

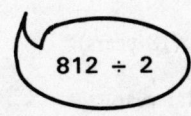

```
    421
  _____
2 ) 842
```

```
    421
  _____
2 ) 842
    8
    ──
    04
    4
    ──
    02
    2
    ──
    0
```

```
   46
  ____
2 ) 812
```

```
   46
  ____
2 ) 812
   8
   ──
   012
    12
   ──
   00
```

The mechanics take over even in an apparently simple task like $812 \div 2$; a commonsense feel for number flies out of the window.

Looking

Task 23) 4669

Look at Table 2.2:

Table 2.2
Task 23) 4669

Students	% choosing	
	Correct answer 203	Most popular wrong answer 23
Secondary students (13/14 years)	46 **38**	40 **35**
TOPS trainees (19–57 years)	39	49
TEC level I (18 years)	40	54
University 1st year design technology (19 years)	53	38

Numbers in **bold type** refer to slower learners

Listening
15-year-old Peter:

'Calculate 23... divided... no, four hundred and sixty... four thousand six hundred and sixty nine divided by 23. 23s into 46... goes two, put the two on top; 23 into 69... goes... three. Answer *23.*

Learning

How do looking and listening help in the diagnosis? Misconcepts arising at school level do matter! The language-laboratory recordings in particular reveal most students attempting to solve tasks by routines, often only partially remembered and partially understood. Well-established correct routines are important to human activities and not least to work in mathematics; they help us to work efficiently and free our minds for the next layer of learning. But these routines should be the result of sound understanding of concepts and this can be brought about by encouraging flexibility of thinking in our students. 'There is more than one way to solve a problem' is an important message in mathematics. Is our student's learning a reflection of our ways of teaching? Do we teach mathematics entirely algorithmically; do we teach for the understanding of concepts but with little practice to reinforce these concepts?

We are encouraged by Z.P. Dienes to make use of the principles of mathematical and perceptual variability, that is to help our students acquire a concept by a variety of approaches via concrete experiences and the more formal experiences with mathematical symbolism.

The pros and cons of 'instrumental understanding' (rules without reasons), 'relational understanding' (knowing what to do and why) and the relevance to teachers and learners is fully discussed by Richard Skemp. You will find this of help in your own thinking of diagnosis and prescription.

The diagnosis

The general diagnosis is *deficiency in a feeling for number* which encompasses:

30 Diagnosis and prescription in the classroom

- Understanding of place value in the decimal system.

> £3251
> I've won three thousand
> *and* two hundred
> *and* fifty
> *and* one pounds?

- Zero as a number and place holder.

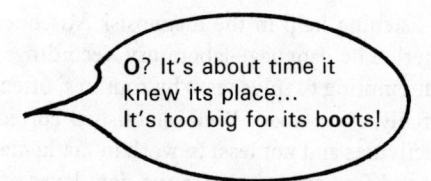

> O? It's about time it
> knew its place...
> It's too big for its b**o**ots!

- The meaning of addition and subtraction (including one as the inverse of the other).

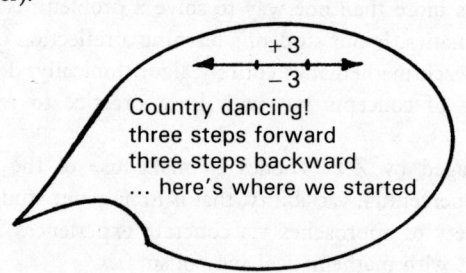

> +3
> −3
> Country dancing!
> three steps forward
> three steps backward
> ... here's where we started

- The meaning of multiplication and division (including one as the inverse of the other).

> 15p
> 15 × 2 → 30 ÷ 2
> First you double my pocket money... then you halve it. I'm no better off!

> Well, at least I can ⊗ and ⊘ with these digits... the Romans couldn't with their fancy ones!

- The meaning of commutativity in addition and multiplication (you need not use the word).

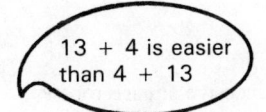

> 13 + 4 is easier than 4 + 13

> 3 × 51 (3 lots of 51) is easier than 51 × 3 (51 lots of 3)

and later in learning ...

> 10 × 0.7 × 3 is easier than 0.7 × 3 × 10

These *can* matter in daily living.

Remember

$$0 \times 7 = 7 \times 0 \text{ is } 0$$
(no lots of 7)   (7 lots of 0)

and $9 \times 0 \times 7 = 0 \times 7 \times 9$ is 0

and even later in learning...

$2\pi r$ is $\pi 2r$
$x^2 y$ is yx^2

Prescription and aftercare

Aim to enhance our students' development of CONCEPTS
  COMPETENCE and CONFIDENCE
    by ourselves being AWARE EXPLICIT and FLEXIBLE.

Any effective treatment should not only cure the misconcept or prevent it arising but should also have a 'knock-on' effect. A successful prescription for multiplication, for example, should also favour the formation of healthy concepts and skills in tasks involving squares and square roots.

A general philosophy for each lesson might be *'give your students a surprise'*.

Investigations with the calculator, for instance, can provide exciting eye-opening experiences with number and provide valuable opportunities for discussion.

Start with a number:
'Homing-in'   What happens when you take succesive square roots?
'Zooming-out'   What happens when you take successive squares?... wow!

Suggestions

Lots of practice:
- Digits and words in place value; stress zero.

H	T	U	Words
	6	3	
5	0	1	
			Twenty nine
			Seven hundred and seven

Fill in the gaps

Pull out the hidden concepts:

 63 is   60 + 3,   i.e. six 10s plus three 1s
501 is 500 + 1,   i.e. five 100s plus one 1.

- Multiplication (and division) by 10, 100, etc.

Use the table at first, moving the digits the appropriate number of places to the left (or right).

- Multiplication by zero.

Use the calculator... what happens?

- Division by zero.

Give them a surprise. Use the calculator. What happens? If your students are ready, use the calculator, start with a number, say 10, and divide 10 each time by a smaller and smaller number...

$$\frac{10}{10} \quad \frac{10}{1} \quad \frac{10}{0.1} \quad \frac{10}{0.01} \quad \frac{10}{0.001} \quad ...$$

what happens? Even mathematicians sometimes have no answer.

- Multiplication with bigger numbers; show thinking behind long $\otimes$.
Pull out the hidden concepts:

Task 407 × 32
is easier as 32 × 407
and 32 × 407 means 30 × 407 + 2 × 407

Quick methods ... or long multiplication
Both should give 13024

NB The concept of number is purely abstract; as soon as we use number with measure we are talking about different physical quantities, e.g. 3 metres is different from 3 elephants.

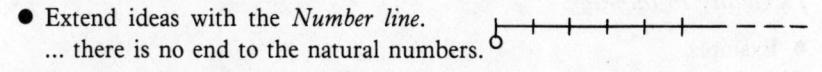

- Extend ideas with the *Number line*.
  ... there is no end to the natural numbers.

- Investigations.
Hand in hand with this kind of practice encourage students to find:
  Patterns, relations, how numbers grow (squares, cubes, etc.)
  Number sequences (Fibonacci, etc.)
  Number patterns with the operators $\otimes$ $\oslash$ $\oplus$ $\ominus$ (remember order).

Some questions for discussion

1. Do we sufficiently emphasize the 'hidden concepts' in number?
2. Do we encourage our students to look for short cuts with tasks such as 41 × 40 and 513 ÷ 27?
3. To what extent do your answers to questions such as these depend on the ability of your students?
4. Do we stretch our more able learners?
5. Do we need to rethink our teaching of the slower learner?

6. Do we encourage too much/too little rote learning? Do we encourage flexibility, generalization for *all* our learners?

Some quotations for discussion

'There is real concern about the lack of basic skills in arithmetic, but efforts to remedy this take the form of more and more sets of questions on fractions and decimals *without identifying the basic number deficiencies that are at the heart of the problem*'. (Our italics)

[*Aspects of Secondary Education in England* (1979) HMI report.]

'The problem of teaching mathematics is the problem of teaching it with *meaning*.' (Our italics)

[*Mathematics 5–11* (1979) A handbook of suggestions. HMSO.]

Diagnostic exercises

Flexibility in thinking

● Example.
Different ways of writing a division task.

Look at this example:

$4284 \div 21$

We can also write this as:

$$21 \overline{)4284}$$

$$\frac{4284}{21}$$

How many 21s in 4284?

Different ways of solving this task:

Short and sweet?

$$21\overline{)4284} = 204$$

Reduction... Factors... Equivalent fractions...

$$\frac{4284}{21} = \frac{7 \times 612}{7 \times 3}$$

$$= \frac{3 \times 204}{3 \times 1}$$

$$= 204$$

```
    204
21)4284
   42
    08
     0
    84
    84
    00
```

If algorithmic do not miss out a step! This may be the problem of the embedded zero!

```
 4284
  -21
 4263
  -21
 4242
  -21
 4221
  -21
  42...
```

Not to be recommended for efficiency but you can build on this level of understanding!

There is a limited number of ways of writing this task when using mathematical symbols but when we set it in a problem-solving situation there are as many contexts as we can find; for example:

- Contexts.

During our 3 week touring holiday of the Highlands we travelled 4284 miles. If we had travelled the same distance each day how far would we have travelled in a day?

A charity raises £4284 for the education of 21 children in India. How much is spent on each child's education?

On average 21 books fit on a library shelf. If the library has a special collection of 4284 books, how many shelves will the collection require?

These tasks could all be put into a multiple-choice format.

- Multiple-choice format (needs to be carefully thought out).

Each 'attractor' must have a sound reason to attract! Good attractors for the example above (4284 ÷ 21) are:

24   240   and   2040   since we have seen students working:

$$21 \overline{)4284} = 24$$

$$21 \overline{)4284} = 240$$

$$21 \overline{)4284} = 2040$$

These would still be valid attractors for the contextual tasks. However, you may find others would be more suitable depending on the context. When designing attractors bear in mind the 'impact perception' that the task may generate. An awareness of this is important if you are to develop your own multiple-choice tasks; the 'e' space can help in this development by collecting alternative attractors.

Guideline exercises

Try all these exercises with your students. Remember they are guideline exercises from which you can construct more of your own. The attractors are designed to reflect students' ways of thinking. Can you see what they are?

Guideline exercise 1

This exercise is specifically about multiplication of smaller whole numbers. How will your students tackle these tasks? Will they demonstrate an understanding of commutativity? What are six lots of zero?

9 × 8 is?
a 17   b 63   c 72   d 98   e

6 × 0 is?
a 60   b 6   c 1   d 0   e

3 × 19 is?
a 22   b 37   c 57   d 60   e

31 × 3 is?
a 90   b 91   c 93   d 313   e

15 × 8 is?
a 158   b 120   c 100   d 80   e

Guideline exercise 2

This exercise is concerned with multiplication of larger whole numbers. In what ways do you expect your students to solve them? How will they cope with multiplication by zero?

60 × 63 is?
a 3680 b 3743 c 3780 d 3843 e

51 × 53 is?
a 2703 b 2603 c 1765 d 408   e

51 × 50 is?
a 255   b 2500 c 2550 d 2601 e

60 × 0 is?
a 600   b 60   c 1   d 0   e

24 × 35 is?
a 210   b 740   c 840   d 1470 e

This can be put in context:
If a glass holds 350 ml of liquid, how many ml will 24 similar glasses hold?
a 14700 b 8400 c 7400 d 2100 e

A trainee is told the wages will be about £30 a week. If this was the exact wage how much will be earned in a 52-week year?
a £156   b £208   c £1560   d £1612   e

A residential home receives £47 per year from the Local Authority to spend on clothes for each child. How much would the matron of a home for 14 children receive to spend on their clothes?
a £658   b £558   c £470   d £235   e

A wine bar has a rack with 25 shelves and each shelf can take 24 bottles of wine. How many bottles of wine can the rack take?
a 480   b 500   c 580   d 600   e

A special car manufacturer produces 41 cars a week in a 40-week production year. How many cars do they produce in their production year?
a 1681   b 1640   c 1600   d 164   e

Guideline exercise 3

This exercise concentrates on division by smaller whole numbers. Watch for the responses to the embedded zero tasks.

$1923 \div 3$ is?
a 31   b 301   c 608   d 641   e

$4 \overline{)804}$ is?
a 21   b 201   c 210   d 2010   e

$4963 \div 7$ is?
a 79   b 709   c 790   d 7090   e

$9 \overline{)3519}$ is?
a 302   b 390   c 391   d 402   e

How many 8s in 856?
a 1070   b 170   c 107   d 17   e

A canteen chef estimates he will serve 490 portions of chips. If there are seven portions to 1 kilogram of chips, how many kilograms are likely to be used?
a 490   b 140   c 70   d 7   e

How many taxis are needed for a party of 31 if a taxi can carry five people?
a 31   b 7   c 6   d 5   e

If I earn £408 in a month how much is this in a week (take one month as four weeks)?
a £1020 b £120 c £102 d £12 e

Five hundred and twenty-five first-class stamps are to be equally shared between five secretaries. How many stamps does each secretary receive?
a 1050 b 150 c 105 d 15 e

My car travelled 384 miles on 8 gallons of petrol. The number of miles the car travelled per gallon was?
a 10 b 40 c 45 d 48 e

Guideline exercise 4

Here we are looking at division by larger whole numbers. What methods will your students use to solve these?

494 ÷ 26 is?
a 9 b 17 c 18 d 19 e

25)2625 is?
a 11 b 15 c 101 d 105 e

2968 ÷ 14 is?
a 212 b 205 c 25 d 20 e

15)4815 is?
a 331 b 321 c 301 d 31 e

4284 ÷ 21 is?
a 24 b 204 c 240 d 2040 e

A hospital ward has a floor space of 1250 square feet. If each patient is expected to use a floor space of 50 square feet, how many patients should the ward take?
a 25 b 24 c 23 d 21 e

One pint of milk is sufficient for 15 cups of tea. How many pints of milk are needed for 3060 cups of tea at a busy railway cafe?
a 24 b 204 c 240 d 2040 e

A secretary sends out 21 letters a day, on average. If she has a stock of 4284 envelopes how many days will the stock last?
a 4284 b 204 c 24 d 21 e

After 27 contributions a charity had collected £513. If each person had contributed the same amount of money they would have each given?
a £17   b £19   c £27   d £513   e

A gardener has 4815 bulbs to plant in fifteen flower beds. How many bulbs are planted in each flower bed if the flower beds each have the same number of bulbs?
a 331   b 321   c 301   d 31   e

CHAPTER 3

SMALL NUMBERS : DECIMAL FRACTIONS

'God created the natural numbers...All else is the work of man.'
(Kronecker)

By small numbers we mean those between 0 and 1. There has probably been more lament by teachers and more discussion on inservice courses over students' inability with both decimal and common fractions than with any other mathematical topic, except possibly negative numbers! Each successive year student teachers return from their first teaching practice with an amazed disbelief at children's antics with fractions.

In this chapter we shall be concerned with the decimal form of fractions, called *decimal fractions,* with which our students can be made readily familiar using the calculator. They are important because they crop up everywhere: man-made activities unfortunately cannot be expressed entirely by using the natural numbers! Decimal fractions are used in measurement of all kinds because they are easy to compare: calculating the speed of a snail in a biology lesson, or a rocket in outer space, and in mathematics squaring a progressively smaller number when exploring the concept of 'limit' in the calculus.

Addition and subtraction appear to present less difficulty for most students provided, of course, that they have been presented with examples and sufficient practice. The tasks can be carried out routinely, the understanding *is not really probed* and so, in general, the misconcepts are not revealed. As with the natural numbers, it is the *multiplication and division* tasks which most clearly expose the misconcepts and even then only certain tasks are effective for diagnosis. This is not to say that there are no difficulties with addition and subtraction: place value will always cause difficulties for those who do not understand. Half of our 10-year-old primary-school children, for example, provided 100 different answers to the task:
add the following numbers 16.36, 1.9. 243.075.

However, we will now start with multiplication.

Diagnosis

Signs and symptoms: multiplication

Looking

Is this what you see when they multiply 'mixed' decimal numbers?

Task 2.4 × 1.3

```
  2.4          2.4
× 1.3        × 1.3
─────        ─────
 24.0         240
  7.2          72
─────        ─────
 31.2         31.2
```
13-year-old school students

Mechanics in action.

Task 37.2 × 100

```
 3    7    2
 3    7    2
 3    7    2
 3    7    2
 3    7    2
 3    7    2
 3    7    2
 3    7    2
 3    7    2
 3    7    2
───  ───  ───
30   70   20
```
17+ craft student (chose to multiply 37.2 by 10... 100 was too difficult)

(120)

Some glimmering of the multiplication process. 'Dot' meant nothing.

28-year-old
TOPS trainee

```
  1.38
  14
─────
  553
 1384
─────
 19.37
```

11 × 9.7

            (x) by integer   (x) by a decimal fraction

            Could do this   ...   but not this

(no understanding of commutativity).

Small numbers: decimal fractions

And when both numbers are small?

```
  0.5        0.3        0.9
× 0.5      × 0.3        0.9
 ─────      ─────      ─────
 00.0       00.0        000
  2.5        0.9         8.1
 ─────      ─────      ─────
  2.5        0.9         8.1
```
13-year-olds

```
  0.4        0.7
× 0.4        0.8
 ─────      ─────
  000        000
   16         56
 ─────      ─────
  01.6       5.6
```

Long multiplication as if with whole numbers and the rule 'keep the points one under the other...' (as in addition and subtraction?).

Listening

14-year-old Andrew:

Task 0.5 x 0.5

'Nought point 5 times nought point 5. First of all you put a line.
0 times 5 is 0. 0 times 0 is 0.
5 times 5 is 25, put the 5 carry the 2.
5 times 0 is 0. So you put down the 2.
Put the point in, 2.5.'

Andrew said he was happy with his answer.

'Keeping the points one under the other' makes the answer bigger than it should be and reinforces the ingrained reaction to multiplication: 'It gets bigger when you multiply'.

The task **0.50 x 0.25** is a good diagnoser of a feeling for number: it is so easily convertible into common fraction form that it readily exposes the understanding. Yet of the many students who attempt this task very few thought of converting it into $\frac{1}{2} \times \frac{1}{4}$. Long multiplication was commonly applied with a range of answers mostly much bigger than 1.

```
0.50 × 0.25 → 1250, 12.5, 1.25, 0.0125
```

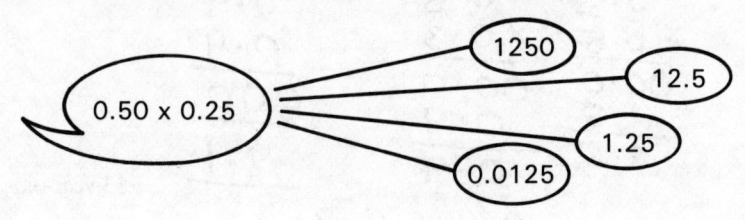

Table 3.1
Task   0.50 × 0.25

Students	% getting correct answer 0.125	Wrong responses (order of popularity)
13+ (mixed ability)	40	12.5 1250 0.0125
14/15 'O' & CSE	56	0.0125 1250 12.5
Craft & technician 17+	45	1.25 12.5 0.0125
TOPS trainees (19–57 years)	46	12.5 1250

Here is one of the few students to whom the task presented no problem (we should of course expect it of this kind of student).

1st year undergraduate in building technology:

'$\frac{1}{2} \times \frac{1}{4}$ equals $\frac{1}{8}$ which is 0.125. *Basic.*'

However 'basic' this task may be, some craft students, on the other hand, worked out 50 × 25 and were happy to have 1250 as the answer; some of these students had no appreciation at all for the decimal point.

'It spoils good numbers, never could see the point of it at school' said one student... and he hoped to be a self-employed electrician!

Craft student:

This student felt that the result should be a large number.

'Fifty times twenty five...it's a large number...twelve whole ones and a half... .'

'Listen' carefully to the way in which this next school student is talking his way through the mechanics of the routine.

13-year-old schoolboy:

' '0' point fifty times '0' point twenty five is... is... um... '0' point fifty... multiplied by '0' point twenty five... um... down nought... fives into a zero goes nothing... five fives's are twenty five carry two... five nuthinks are nuthink... twos into nuthink... five two's are ten carry one... two nuthinks nuthink... add it up... nought nought six... six hundred... '0' point fifty times '0' point twenty five... is... um... nought point one two five.'

He arrived with the correct solution; had he much feeling for number?

Learning

Teachers' expectations of 13+ students on the task **0.50 x 0.25** were interesting. They underestimated the difficulty but more significantly most of them thought that the errors were more likely to be of the sort 0.0125 rather than errors bigger than 1, i.e. students were credited with a more sophisticated misconcept.

Table 3.1 highlights the nature of the misconcepts. What emerges quite clearly is that the same misconcepts can persist not only through a learner's school life but through continuing education and into adult life. The adults on TOPS courses, for example, span quite a wide age range from the 19-year-old who 'never could do maths at school' to the older adult who needs to be reminded.

One of our 28-year-old trainees had left school at 17 with no formal qualification in mathematics. She was very confused with decimal fractions:

'I don't know what 0.9 means... I'm absolutely useless when it comes to a dot'.

Most students appear then to solve tasks involving multiplication of decimal fractions in the same rote way in which they use long multiplication for the natural numbers. Very few demonstrate any commonsense appreciation for the size of their number solution.

If we 'look and listen' again and 'home-in' to certain tasks we may learn short cuts to diagnosis.

Looking

Consider for example two tasks which look so alike.
Here is what could happen when we set a *written* exercise:

A learner

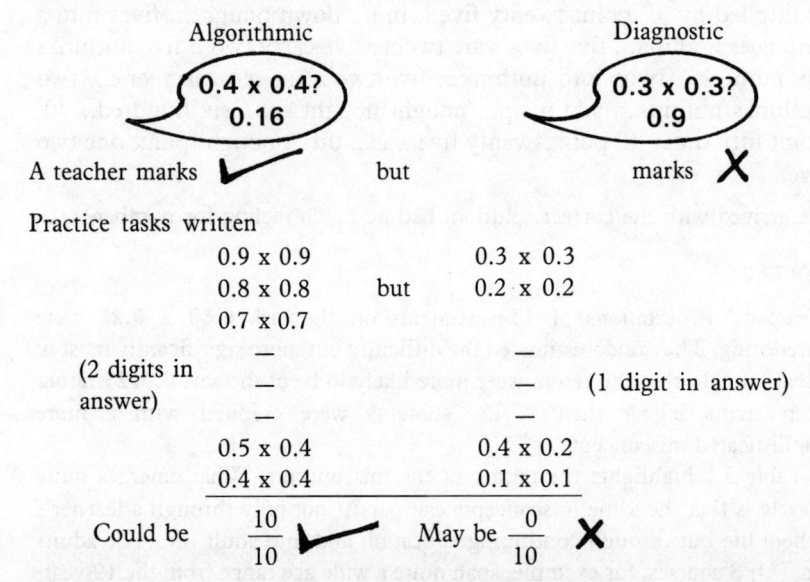

                 Algorithmic   Diagnostic

                   0.4 x 0.4?   0.3 x 0.3?
                    0.16   0.9

A teacher marks ✓   but   marks ✗

Practice tasks written

0.9 x 0.9		0.3 x 0.3
0.8 x 0.8	but	0.2 x 0.2
0.7 x 0.7		

(2 digits in answer)   (1 digit in answer)

                 0.5 x 0.4   0.4 x 0.2
                 0.4 x 0.4   0.1 x 0.1

Could be $\dfrac{10}{10}$ ✓   May be $\dfrac{0}{10}$ ✗

It is clear that if we give our students written practice tasks then those of the kind 0.3 x 0.3 behave as better diagnosers than those of the kind 0.4 x 0.4. Can you see how important it is to think carefully of the tasks we set? The most effective diagnosis of course includes listening.

Listening

With the same tasks
13-year-old Micaela:

'Nought point 9 times nought point 9.
So what do I do first, 9 times 9 is 81, then put a point before... after... before the 81 and put nought nought which makes the answer *nought nought point eighty one.*'

'Nought point 7 times nought point 7. 7 times 7 is 49, put the point down and nought before and the answer is *nought nought point forty nine.*'

'Nought point 5 times nought point 5 equals 5 times 5 is 25, put the point before it and put 2 noughts down. So the answer is *nought nought point twenty five.*'

'Nought point 4 times nought point 4 is 4 times 4 is 16, put the point before the 16, put 2 noughts down, so the answer is *nought nought point sixteen.*'

In the *written* situation we would mark these as correct, maybe pointing out that one zero before the decimal point is sufficient, and assume that the student understood.
But here we can see the role of *verbalizing*. The way this student stated the answers indicates the way she was thinking. Listen in again...

'Nought point 3 times nought point 3. 3 times 3 is 9, put the point down before the 9 and put nought nought. So the answer is *nought nought point nine.*'

'And the last one, nought point 1 times nought point 1. 1 times 1 is 1, put a nought, put the point before that, put 2 noughts down, *nought nought point one.*'

And these tasks of course expose the misconcept. This student was applying a rule mechanically: multiplying the digits as if they were whole numbers and then simply sticking in a decimal point in front of the answer. She showed some idea of the need to count the number of decimal places by the 'nought nought' routine.

Learning

The response of 0.9 to the task **0.3 x 0.3**, which is so popular with all learners represents an automatic/rote/trigger response. Even of her able 'O' level pupils one teacher said this:

'They tend to give an answer straight off – they say, . 3 times .3 equals .9 without thinking. They do it in exactly the same way as they say 3 times 4 equals 12 – they know that's so and they just as much know .3 times .3 is .9 and, in fact, when I tested my 13-year-old set, nearly half of them said this.'

Although some teachers anticipated this kind of response from some of their pupils they did significantly underestimate its popularity.

Awareness of the nature of the misconcepts which arise and the extent to which they arise is necessary for our teaching to be effective.

'I have to admit that, I had not really considered this as a separate problem from 'mixed decimals', that is, it had not dawned on me that 1.6 x 1.6 has less attendant difficulties than 0.6 x 0.6. Occasionally I have been called to a desk and been asked: "Where have I gone wrong?" when a correct answer has been obtained. Further questioning has produced: "I've multiplied, but I've got a smaller answer than the number I started with". This has given rise to a short class discussion on what happens when we multiply by numbers less than one, but I have never set out to teach this explicitly in the context of *decimal* fractions.'

Awareness by the teacher is the first step and is crucial but the misconcept has a firm hold as another teacher expresses:

'It gets bigger when you multiply' gets stuck in their mind – it's like bindweed, you pull it up and there it is next week just as bad... .'

This teacher said that it seemed possible to get rid of a first misconcept only by using the concept again and again in different ways, keeping it in the front of students' minds all the time.

Signs and symptoms: division

Looking

Is this what you see?

Task $7.8 \div 3$

$$3 \overline{\smash{\big)}\, 78}^{218}   3 \overline{\smash{\big)}\, 7.8}^{2.6}$$

13-year-old

Confusion...?

Small numbers: decimal fractions 49

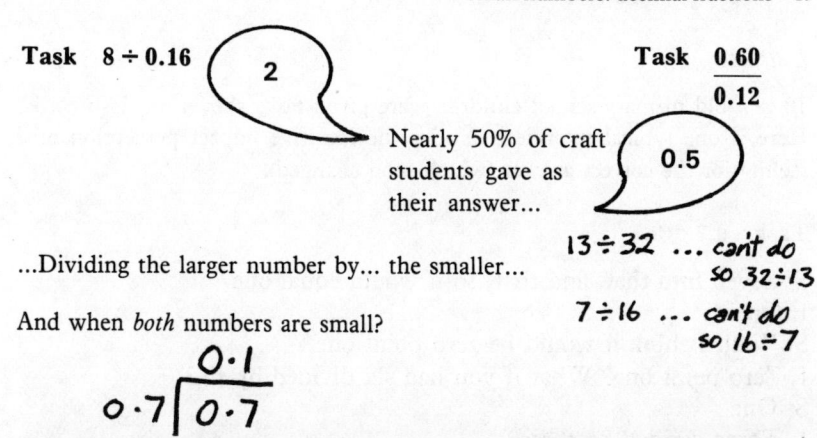

Task $8 \div 0.16$ *(speech bubble: 2)*   Task $\dfrac{0.60}{0.12}$ *(speech bubble: 0.5)*

Nearly 50% of craft students gave as their answer...

...Dividing the larger number by... the smaller...

$13 \div 32$ ... can't do
so $32 \div 13$

And when *both* numbers are small?

$7 \div 16$ ... can't do
so $16 \div 7$

$$0.7 \overline{\smash{)}0.7}^{\,0.1}$$

The misconcepts arising from division appear to stem from the same kind of thinking as with multiplication. The small numbers are treated as if they are whole numbers and a point is 'stuck' in the seemingly most appropriate position.

For student responses to the task **0.7 ÷ 0.7** see Table 3.2.

Table 3.2
Task 0.7 ÷ 0.7

	% choosing	
Students	Correct answer	Most popular wrong response
	1	0.1
Primary		
10+ (mixed ability)	49	24
Secondary		
12+ (mixed ability)	32	24
13+ (mixed ability)	29 **6**	42 **57**
14+ 'O' & CSE	62	24
TOPS trainees	27	46

Numbers in **bold type** refer to slower learners

NB 0.1 was the most popular wrong response by students at school and adults.

10-year-old junior pupils, who appeared at the time to be practising tasks of this kind, did better than those at second- and third- year secondary level.

Listening

10-year-old primary-school children were given tasks shown on flash cards. Here is one typical interview. Notice the intuitive impact perception of a feeling for the correct answer (which soon changed).

Task   0.7 ÷ 0.7

S: Seven into that, and that, so it would equal one.
I: One?
S: Well. I think it would be zero point one.
I: Zero point one? What if you had six divided by six?
S: One.
I: Three divided by three?
S: One.
I: What would one half divided by half be?
S: A half! Point five!
I: A quarter divided by a quarter?
S: A quarter!

Adult (professional occupation):

'Nought point seven divided by nought point seven... point one of course... because seven divided by seven is one... so point one.'

Misconcepts which start early on are difficult to dislodge and may cause dislike of maths through a lifetime for many kinds of people. As one teacher working with TOPS trainees said: 'multiplication and division of decimal fractions proved to be a very weak area, with very few understanding fully the significance of the decimal point. The integer part of the calculation posed no real problems but when multiplying or dividing decimal fractions the methods became very erratic and illogical'.

Learning

Teachers' predictions of their 13-year-old students' performance on the task **0.7 ÷ 0.7** were considerably higher than in reality: 48% compared with 29%.

There appears to be a chasm between working with the natural numbers and decimal fractions. Students are clearly trying to relate the new learning associated with small numbers to their experience with the natural numbers:

Small numbers: decimal fractions 51

'It gets smaller when you divide... '. Lack of feeling for number just about sums up the misconcepts, and the decimal point for many students is an unnecessary complication!

As one teacher summarized the problems:

'One of the main problems is just the decimal point. It looks like a little dot, very insignificant, and is regarded as being very insignificant. Another thing they find difficult is the whole idea of the size of numbers and the idea of place value. It seems that this concept of place value is one of the most difficult to understand.'

The 'knock-on effect': squares and square roots

Misconcepts arising from multiplication and division will, of course, seed and flourish into work with squares and square roots.

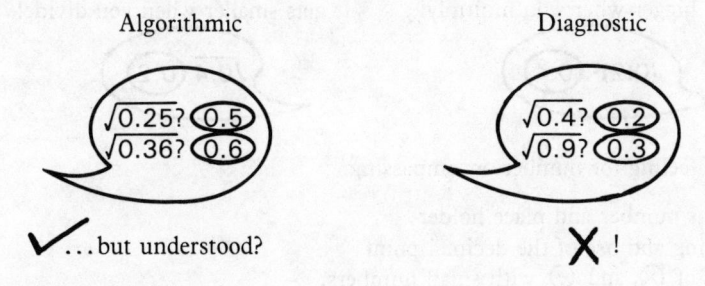

Algorithmic

$\sqrt{0.25}?\ \boxed{0.5}$
$\sqrt{0.36}?\ \boxed{0.6}$

✓ ...but understood?

Diagnostic

$\sqrt{0.4}?\ \boxed{0.2}$
$\sqrt{0.9}?\ \boxed{0.3}$

✗ !

The following recordings are typical of many students:

$\sqrt{0.25}$ 'Square root of nought point 25 is... nought point 5.'
$\sqrt{0.9}$ 'Square root of nought point 9 is... nought point 3... because point 3 times point 3 is point 9.'

Does treatment work?

What then can we do about misconcepts arising with small numbers? The encouraging news is that treatment in the form of explicit teaching does seem to work. As a result of the interest expressed by many teachers, a small in-depth study was carried out in which teachers designed for their 13+ students their own programmes of explicit teaching of the concepts and skills relating to numbers 'less than one'. During the course of the year there was an increase in facility value of 20% for the **0.4 x 0.4** task, 15% for **0.3 x 0.3** and 8% for **0.7 ÷ 0.7**.

One teacher reported that the 13-year-old pupils with whom she had been involved in diagnosis followed by explicit teaching were the year after taught by the headmaster. He stated that he could spot with relative ease those concepts and skills involved in her 'treatment' and quoted as an example the pupils' greatly improved performance and understanding of 'numbers less than one'.

The diagnosis

$0.3 \times 0.3 \; \boxed{0.9}$ $\qquad\qquad \dfrac{0.7}{0.7} \; \boxed{0.1}$

The 'Bindweed' misconcept!

It gets bigger when you multiply! It gets smaller when you divide!

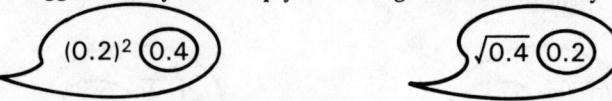

$(0.2)^2 \; \boxed{0.4}$ $\qquad\qquad \sqrt{0.4} \; \boxed{0.2}$

Lack of feeling for number encompassing:

Zero as number and place holder
Meaning and use of the decimal point
Effect of ⊗ and ⊕ with small numbers.

Prescription and aftercare

Aim for students CONCEPTS COMPETENCE and CONFIDENCE
for ourselves AWARE   EXPLICIT   and FLEXIBLE.

Suggestions

Give short sharp surprises:

- Get students to use the calculator/microcomputer to

    ⊗ small numbers... what happens?

    ⊕ small numbers... what happens?

Small numbers: decimal fractions 53

(✓) small numbers... what happens?

( )² small numbers... what happens?

Discuss look, listen and learn with them.

The calculator is an invaluable aid in demonstrating the *generalization* that is part of the power of mathematics. It opens up opportunities for discussing the relative size of numbers:

e.g. *part* of (part of a whole one) is an even *smaller* part of the whole one

(0.2 x 0.3 = 0.06)

part divided by (part of a whole one) is an even *bigger* part of the whole one

($\frac{0.3}{0.6}$ = 0.5)

Give another surprise:

● Take any number, say 10, and divide it by successively smaller numbers, say 10, 1, $\frac{1}{10}$, $\frac{1}{100}$, $\frac{1}{1000}$ and so on... . What happens? Repeat dividing by successively bigger numbers. *Talk* about space, the heavens, atoms...

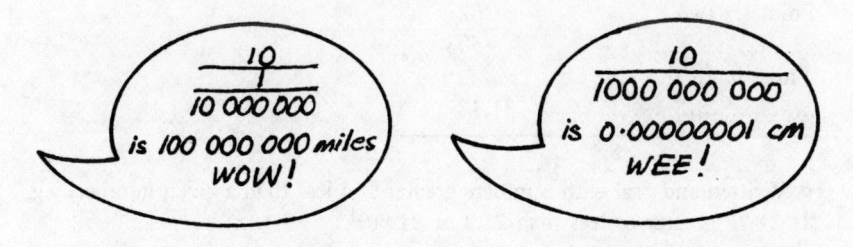

$\frac{10}{\frac{1}{10\,000\,000}}$ is 100 000 000 miles WOW!

$\frac{10}{1000\,000\,000}$ is 0·00000001 cm WEE!

How far away is the sun?   How small can an atom get?

● And of course don't forget to extend the ideas of place value developed with the natural numbers to tenths, hundredths and thousandths.

1 and numbers >1			Decimal	Parts of a whole 1		
H	T	U	point	t	h	th
			•	3	0	
		0	•	5		
	1	2	•	2	4	5

Decimal system

Use the table to establish:

That 1 and numbers greater than 1 are to the left of the decimal point, numbers smaller than 1 to the right

That .3 is the same as $\frac{3}{10}$ that .30 is the same as $\frac{30}{100}$ and that all of these are the same as each other

The role of zero, e.g. 0.3 is the same as .3 is the same as .30, or even 0.300.

Ask questions like 'What is the value of the 4 in 0.74?' or 'What is the value of the 5 in 0.05?' The table will help to give the answer 'meaning'.

Lots of practice:

- In filling in the gaps and naming the numbers.

Words	Numbers
Point six two	.62
....	.503
Three point five six	....
....	31.42

- Written and oral with numbers greater and less than a given number, e.g. 'Is 21.73 greater or less than 21.8 or 21.80?'.

- Emphasize 'hidden' concepts with written and oral examples:

0.35 means 0.3 + 0.05 or $\frac{3}{10} + \frac{5}{100}$

2.57 means 2 + 0.5 + 0.07 or $2 + \frac{5}{10} + \frac{7}{100}$

Small numbers: decimal fractions

- Use the number line

```
├─────┼─────┼─────┼─────┤...
0    .25   .5   .75    1
    or 25/100  or 5/10  or 75/100
```

to show

.5 is half-way between 0 and 1
.25 is half-way between 0 and .5
and half-way between .2 and .3

.75 is three-quarters-way between 0 and 1
half-way between .5 and 1
and half-way between .7 and .8.

And this should help students toward an understanding with common fractions.

We could pull together some of these ideas in aiming for *flexibility*. 'There is more than one way to solve a problem' needs to be reiterated. Yet as we know great skill is then needed in our teaching to avoid confusion by the student.

The strength of our understanding lies in the flexibility with which we can use our knowledge in many varied contexts. Here is one example to try with your older students:

$8 \div 0.16$ is?
If $0.16x = 8$, then x is?

Given that resistance = $\dfrac{\text{potential difference}}{\text{current}}$

an appliance carrying a current of 0.16 amperes with potential difference of 8 volts has a resistance of?

Some questions for discussion

1. Tasks involving computation with small numbers may seem elementary, but in fact they require students to grasp some quite complex mathematical concepts. What are these concepts and how best can you help your students to acquire them?

2. 'One of the main problems that students have is the decimal point. It looks

like a little dot, very insignificant.' Do you find this? How best can you help your students to understand the concepts associated with place value?
3. 'Current teaching underplays mental arithmetic' *(Mathematics* 5–11). Do you agree? Do you have mental arithmetic sessions with your students? Regularly? What do you see as the role of mental arithmetic? Has it any drawbacks?
4. 'To develop an appreciation of mathematical pattern and the ability to identify relations is one general aim of teaching maths' *(Mathematics* 5–11). Should this be the aim for all your students? How would you try to achieve it using decimal fractions?

Guideline exercises

Guideline exercise 1

This exercise is specifically about multiplication of numbers less than one. How will your students tackle these tasks? (Notice that they have been written horizontally.)

0.2 x 0.1 is?
a 0.0002   b 0.02   c 0.2   d 2   e

0.4 x 0.5 is?
a 20   b 2.0   c 0.2   d 0.02   e

0.3 x 0.2 is?
a 6   b 0.6   c 0.5   d 0.06   e

0.20 x 0.25 is?
a 0.05   b 0.5   c 5   d 5000   e

0.3 x 0.15 is?
a 45   b 4.5   c 0.45   d 0.045   e

The cost of 0.64 kg of steak at £4.25 per kg is?
a £27.20   b £6.97   c £2.81   d £2.72   e

The area in square metres of a carpet tile 0.25 m by 0.25 m is?
a 6250   b 62.5   c 0.625   d 0.0625   e

Small numbers: decimal fractions 57

If the probability that an 'average middle-aged man' will be alive in 30 years is 0.2 and the probability that an 'average middle-aged woman' will be alive in 30 years is 0.3 then the probability that they will both be alive in 30 years is?
a 0.06   b 0.1   c 0.5   d 0.6   e

Guideline exercise 2

This exercise concentrates on division involving numbers less than one. Once again the tasks have been written horizontally.

0.5 ÷ 0.5 is?
a 0   b 0.1   c 0.25   d 1   e

0.8 ÷ 0.4 is?
a 5   b 2   c 0.5   d 0.2   e

0.3 ÷ 0.6 is?
a 0.05   b 0.5   c 0.2   d 2   e

0.9 ÷ 0.03 is?
a 0.3   b 3   c 30   d 300   e

5.4 ÷ 0.9 is?
a 60   b 6   c 0.6   d 0.06   e

How many bottles of wine containing 0.7 litres can be filled from a barrel containing 5.6 litres?
a 80   b 8   c 0.8   d 0.08   e

How many pieces of ribbon, each 0.08 m long, can be cut from a length of 0.8 m?
a 100   b 10   c 0.1   d 0.01   e

The area of a tile is 0.09 square metres. If the length of the tile is 0.3 metres what is its width?
a 0.03 m   b 0.3 m   c 3 m   d 30 m   e

Guideline exercise 3

This exercise concentrates on addition and subtraction tasks for small numbers.

0.5 + 0.04 is?
a 0.09   b 0.9   c 0.45   d 0.54   e

0.70 + 0.95 is?
a 165   b 16.5   c 1.65   d 0.0165   e

0.2 − 0.15 is?
a 0.05   b 0.13   c 0.15   d 0.5

0.94 − 0.75 is?
a 0.29   b 0.21   c 0.19   d 0.11   e

0.98 − (0.5 + 0.29) is?
a 0.19   b 0.21   c 0.71   d 0.77   e

CHAPTER 4

SMALL NUMBERS: COMMON FRACTIONS

'I only took the regular course... the different branches of Arithmetic — Ambition, Distraction, Uglification and Derision... .'

(Lewis Carroll)

This chapter is concerned with exploring the understanding of number via common fractions, those numbers which we use in everyday language: one quarter of a pound, $\frac{3}{4}$ of an hour and so on. Do your students understand the notion of a common fraction or are they like this boy? !

Normally I cut my pie into 2 ... *but when I'm hungry I cut it into 4*

The discussion has been separated from the previous chapter for clarity but nevertheless in our teaching we should use common fractions and decimal fractions together whenever possible to emphasize that they are different forms of the same thing.

11-year-old Martin seems to understand this:

'Well, if its .81 it will be $\frac{81}{100}$

.49 its $\frac{49}{100}$

.25 is $\frac{25}{100}$

.09 ... well that would be $\frac{9}{100}$, which doesn't cancel so I'd be left with $\frac{9}{100}$'.

Decimal fractions are common fractions using tenths, hundredths, etc. They tend to be used more in calculation since they are more easily compared. Even so, it appears to be the sad reality that few of our students have a sufficiently developed feeling for number to enable them to switch naturally from decimal form to fractional form and vice versa as we saw with the task **0.50 x 0.25** in Chapter 3.

Unlike work with decimal fractions, multiplication of common fractions *apparently* causes students little trouble. 'Multiplying tops, multiplying bottoms' gives the right answer and appears the obvious routine to use. It is one of the few instances in maths, where, superficially, what seems an obvious procedure is also a correct procedure. So correct answers to tasks involving ⊗ give little guidance to our students' understanding of common fractions.

Diagnosis

Let's look at some of our students' work.

Signs and symptoms: addition and subtraction

Looking

Are these the ways in which they add common fractions or subtract them? If so, it leads to trouble later with algebraic fractions. Even engineering undergraduates have trouble in adding quantities of the kind $\frac{a}{2} + \frac{b}{5}$. What do you think they write?

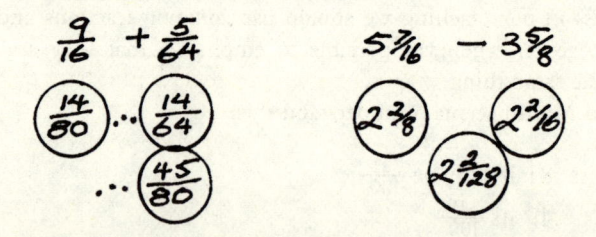

$$\frac{9}{16} + \frac{5}{64}$$

$$\left(\frac{14}{80}\right) \cdots \left(\frac{14}{64}\right)$$

$$\cdots \left(\frac{45}{80}\right)$$

$$5\frac{7}{16} - 3\frac{5}{8}$$

$$\left(2\frac{2}{8}\right) \quad \left(2\frac{2}{16}\right)$$

$$\left(2\frac{2}{128}\right)$$

We set the following tasks to students at school and college.

Small numbers: common fractions 61

Task $\frac{9}{16} + \frac{5}{64}$

44% correct (38% chose $\frac{14}{80}$)	59% correct (27% chose $\frac{14}{80}$)
13+ (mixed ability)	14+ ('O' and CSE)

Task $\frac{3}{16} + \frac{5}{64}$

58% correct (26% chose $\frac{8}{80}$)

Craft/technician 17+

The most popular incorrect response is 'adding tops, adding bottoms'.

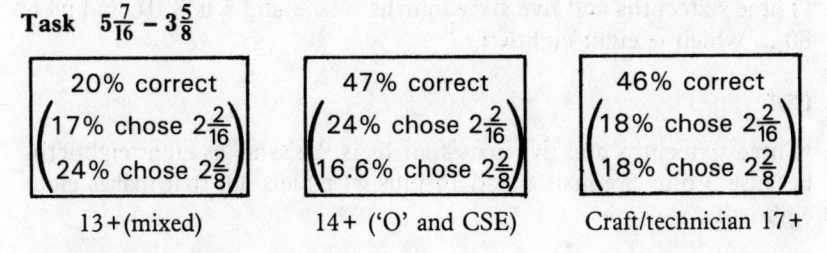

Task $5\frac{7}{16} - 3\frac{5}{8}$

20% correct $\left(\begin{array}{c}17\% \text{ chose } 2\frac{2}{16}\\ 24\% \text{ chose } 2\frac{2}{8}\end{array}\right)$	47% correct $\left(\begin{array}{c}24\% \text{ chose } 2\frac{2}{16}\\ 16.6\% \text{ chose } 2\frac{2}{8}\end{array}\right)$	46% correct $\left(\begin{array}{c}18\% \text{ chose } 2\frac{2}{16}\\ 18\% \text{ chose } 2\frac{2}{8}\end{array}\right)$
13+(mixed)	14+ ('O' and CSE)	Craft/technician 17+

You see that:
- The favourite incorrect responses were:
  'subtract the whole numbers then subtract tops and subtract bottoms';
  'subtract the whole numbers then subtract tops and choose the first bottom'.
  These solutions are of course a purely mechanical application of a rule... there was no feeling that $\frac{5}{8}$ could possibly be larger than $\frac{7}{16}$.
- The misconcept of the 13/14-year-old at school carries through into college life at 17+.

There was a variety of other incorrect responses which reflected partially remembered rules about 'denominators' and 'numerators'. These very words seem to conjure up for our students mild forms of mental torture.

We have included powers of two in our studies since 1970 because industry and college engineering courses at that time emphasized the need for these fractions to be met at school. Twelve years later this need still exists as the Cockcroft Report points out.

It is clear from these few examples that many students are mechanically applying rules with little or no understanding of the concepts.

Let's probe further and listen in to their thinking.

Listening

Here are two thoroughly confident 13-year-olds in the language laboratory: Notice their 'triggered' responses.

Task $\frac{3}{16} + \frac{5}{64}$

Boy

'Three sixteenths and five sixty-fourths is ... 3 and 5 is 8, 16 and 64 is 80 ... which is eight eightieths.'

Girl

'Three sixteenths and five sixty-fourths is the same as eight eightieths because 3 plus 5 equals 8 and 16 plus 64 equals 80; that makes eight eightieths.'

Here is a 12-year-old being interviewed:

Task $\frac{3}{8} + \frac{5}{16}$

S: 5 add 3 equals 8 and 8 over 16 is 24 and I divide into it and it becomes ... I divide by 2 ... and it's 4 over 12 and I can divide into that again to make it 1 over 3.
I: Would you tell me again what you did.
S: 5 add 3.
I: That gave you 8, then what did you do to the bottoms ... the 8, the 16?
S: Added them as well.

This boy was quite confident in his routine; he added tops and bottoms and got $\frac{8}{24}$. He then showed a grasp of equivalent fractions (or was it just another routine?); $\frac{4}{12}$ and finally $\frac{1}{3}$. He clearly enjoyed working with numbers.

17+ craft students

Task $\frac{3}{16} + \frac{5}{64}$

'I've done these ones but not sure. Can't remember really ... add top two and bottom two ... 3 and 5, 16 and 64 equals $\frac{8}{80}$.'
'... adding 3 and 5 ... making 8 ... 16 and 64... 80 ... $\frac{8}{80}$.'
But all is not lost
S: Try to get these two to same base ... 64 ain't it ...?
I: Yes ... do it ...
S: 4 times that ... so multiply 3, 4 times ... 12 over 64 plus 5 over 64 is 17 over 64.

And this one

'$\frac{3}{16}$ is $\frac{6}{32}$, $\frac{12}{64}$ ... $\frac{12}{64} + \frac{5}{64}$ is $\frac{17}{64}$'

Equivalent fractions were doing well.
What have we learned so far from our looking and listening?

Learning

It is clear that for the purpose of diagnosis it is addition and subtraction which explore the understanding.

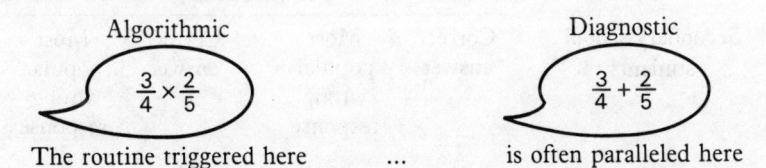

Algorithmic $\quad\quad\quad\quad\quad\quad$ Diagnostic

$\frac{3}{4} \times \frac{2}{5}$ $\quad\quad\quad\quad\quad\quad$ $\frac{3}{4} + \frac{2}{5}$

The routine triggered here $\quad$ ... $\quad$ is often paralleled here

Our students think 'If common fractions can be multiplied using the rule ... multiply tops, multiply bottoms, ... why can't they be added using the rule ... add tops, add bottoms...?'
It seems logical to them but they are using rules without meaning.

No discussion on common fractions is complete without reference to division ... the ultimate in mystical processes! Division of common fractions will show up the weakness in the understanding of multiplication, and that one is the inverse of the other. Nevertheless it is a process amenable to rote and to the application of the age-old rule 'upside down and multiply'. Many a small primary-school child will never forget!

Common fractions are so much a part of our everyday vocabulary that we might expect students to be able to cope much better with fractions in real-life contexts. We discuss this more fully in Chapter 10 but give, briefly, one example here (Table 4.1).

Table 4.1 Effect on students' success rate of putting fractions into an applied context.

	$1 - (\frac{1}{3} + \frac{1}{4} + \frac{1}{8})$		Real-life context	
	% choosing			
Secondary-school students	Correct answer $\frac{7}{24}$	Most popular wrong response $\frac{3}{15}$	Correct answer $\frac{7}{24}$	Most popular wrong response $\frac{3}{15}$
13+ (mixed ability)	27	23	40	15
13+ (slower learners)	9	48	26	34
14+ 'O' & CSE	56	18	60	16

There is a significant increase in success rate for the 13-year-olds across the ability range. The evidence collected in Chapter 10 seems to suggest that, while context can be helpful, the nature of the context is crucial. This is where the teacher's judgement of what is *relevant* and accessible for the student is so important.

Exploring the understanding further

The real evidence of understanding is probably the flexibility to convert from one form of a fraction into another, i.e. from common into decimal form and vice versa.

Table 4.2

Task $\frac{7}{16}$ correct to two decimal places is?

Students	% choosing		
	Correct answer	Most popular wrong response	Other response
	0.44	0.43	7.16
13+ (mixed ability)	13	21	35
14+ ('O' & CSE)	31	27	
Craft 17+	24	31	
Technician 17+	35	46	
TOPS trainees (19–57 years)	21	29	15

Looking

Learners, on the whole, even the more able 'O' level students do not find this conversion at all easy as is evident from Table 4.2 and the incorrect answers they chose. The task was originally designed in this way, i.e. correct to two decimal places (see Table 4.2) because of its direct relevance to the vocational training of the student on the workshop floor. A cost-conscious employer is concerned that fabrication trainees, for example, are efficient with their use of materials and machines.

How do younger students cope with a similar task?

Task What is $\frac{3}{8}$ as a decimal fraction?

10% correct	10% correct
(32% chose 3.8)	(37% chose 3.8)
10-year-olds	12-year-olds

Listening

A 10-year-old student:

Task Change $\frac{3}{8}$ to a decimal

S: Reads aloud, repeats... 'Mmmmm...Hmmmm'.
I: Have you done any of these?
S: Yes, I think so.
I: Do you remember what to do? What would one half be as a decimal?
S: A half...mmmm...point five.
I: A quarter?
S: Point twenty five.
I: Did you learn a way to do these? A method?
S: No, just changed them.
I: You didn't learn $\frac{3}{8}$ then?
S: No, just halves, quarters, one eighths.
I: Do you know what one eighth is?
S: If you know what one is you can find three.
I: Good, that's right.
S: So... one eighth is a ... half five...Oh No!

Craft students 17+:

Task $\frac{7}{16}$ correct to two decimal places is?

I: You're meeting decimals and fractions in work. Do you know how to do this one ?
S: Think you multiply it somehow... not sure.
I: Multiply...what by what... ?
S: Top by bottom...I mean top into bottom.
I: You mean divide ?
S: Yeah.
I: You divide 7 by 16... .

The student spent some time on this but did not get very far.

Recordings of students solving this task show that they automatically attempt long division except for those pupils and older students who persist in

Small numbers: common fractions   67

believing that 'you cannot divide a smaller number by a larger'. Many of the students who successfully carry out the long division then fail to understand the meaning of 'correct to two decimal places'.

Learning
Students, including teacher trainees, appear not to see $\frac{7}{16}$ as in any way related to halves, quarters or eighths: one student only was triggered to say the magic words 'that's almost one half'. Again, teachers' expectations of students' performance on this task, that 41% of their 13-year-old mixed-ability classes would do it correctly, was much higher than the reality of 13%, with 8% for the slower learner. Adults on TOPS courses attempted this task and 21% of them solved it correctly.

The diagnosis
● Rigidity of thinking.
Looking almost always for algorithmic solutions betrays a lack of feeling for number and a lack of confidence in working with common fractions which are still part of our everyday vocabulary. It's almost as if learners think that commonsense must fly out of the window as soon as 'mathematics' takes over. The diagnosis is therefore essentially as for the natural numbers and decimal fractions.

Prescription and aftercare

Aim In our treatment as practitioners we must therefore be:
    AWARE   of the real nature of our students' misconcepts
    EXPLICIT   in the teaching of the associated concepts and skills
    FLEXIBLE   in ways of presenting the material.

Aim for our students is to promote their development of
    CONCEPTS
    COMPETENCE in application of concepts
    CONFIDENCE   in working with concepts (and as a pay-off)
    FLEXIBILITY/
    TRANSFER   of the concept to different situations.

What then is the particular treatment for 'common fraction delirium'? A full discussion would almost be a monograph in its own right. What follows are some suggestions of models and algorithms and in Chapter 12 there is more detailed discussion on possible activities.

Common fraction models and algorithms

Most teachers wish to provide their students with working models which give an intuitive feeling for their work with common fractions. In these suggestions each model is followed by an appropriate algorithm. It is hoped that a model will give understanding but for efficient working we cannot always go back to first principles.

An algorithm based on understanding will be better remembered and will not only facilitate efficient and speedy working but should act as a base for future conceptual learning. (More mathematically able students have said that they initially worked algorithmically with fractions until later the rules 'clicked'.)

To help establish a concept students could model first and use the algorithm as a check.

- Start by using their experience of decimal fractions...the 'tenths'.

$\pm$ $\frac{5}{10} + \frac{3}{10}$ is $\frac{8}{10}$; $\frac{5}{10} - \frac{3}{10}$ is $\frac{2}{10}$ (you do not subtract the bottoms)

$\otimes$ $\frac{3}{10} \times \frac{3}{10}$ is $\frac{9}{100}$ or .09 ... check with their knowledge of 0.3 × 0.3

Common fraction model

Part of a whole, equivalent fractions.

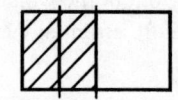

Piaget stresses the importance of children appreciating that fractions are parts of a whole which can be separated and reassembled to form the same whole. Folding and cutting are good activities for this and can also lead to exploring equivalent fractions, i.e. $\frac{1}{2}$ is equivalent to $\frac{2}{4}$ or $\frac{4}{8}$, $\frac{3}{6}$, $\frac{5}{10}$ and so on. And the understanding of equivalent fractions is critical to an understanding of work with fractions.

Common fraction algorithm

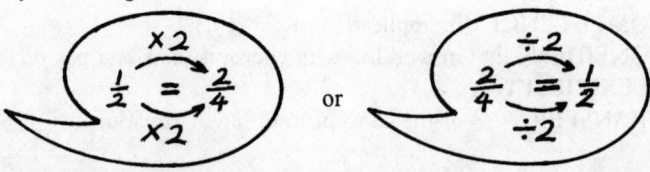

$\frac{1}{2} \xrightarrow{\times 2} \frac{2}{4}$ or $\frac{2}{4} \xrightarrow{\div 2} \frac{1}{2}$

To make an equivalent fraction:
'Multiply top and bottom of fraction by the same number'
or
'Divide top and bottom of fraction by the same number'

Small numbers: common fractions 69

Addition model

It is important to realize that when adding or subtracting 'parts of a whole', *the whole* in each case *must be the same*.

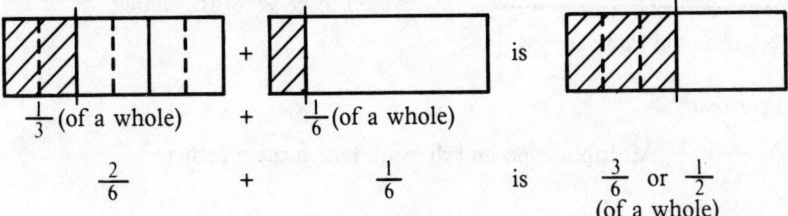

$\frac{1}{3}$ (of a whole) + $\frac{1}{6}$ (of a whole) is

$\frac{2}{6}$ + $\frac{1}{6}$ is $\frac{3}{6}$ or $\frac{1}{2}$ (of a whole)

Addition algorithm

Only fractions of the same kind can be added: the bottom number tells us what kind of fractions (i.e. sixths or tenths), the top one how many of that kind we have:

$\frac{1}{3} + \frac{1}{6}$ must all be sixths.
We need equivalent fractions:
$\frac{2}{6} + \frac{1}{6}$ is $\frac{3}{6}$ or $\frac{1}{2}$.

e.g. $\frac{2}{5} + \frac{1}{4}$ Change to twentieths
$= \frac{8}{20} + \frac{5}{20}$ Add like fractions
$= \frac{13}{20}$.

Calculator (or microcomputer) algorithm

The calculator can also help to give a short sharp surprise to the 'add the tops and bottoms' brigade!

$\frac{2}{5} + \frac{1}{4}$

A calculator *under your instructions* obeys the rules for the order of operations, i.e. 'division precedes addition'.

0.4 + 0.25 is 0.65 ... can they convert into $\frac{65}{100}$ ... to $\frac{13}{20}$ as the simplest equivalent fraction?

Multiplication (and division) model

It is most helpful to replace ⊗ with 'of' and to realize that 'of' means ⊗ only.

$\frac{2}{3} \times \frac{1}{4}$
$\frac{2}{3}$ of ($\frac{1}{4}$ of a whole one)

is $\frac{2}{12}$ (of a whole one)
or $\frac{1}{6}$ (of a whole one)

Compare the model to show division as the inverse of multiplication.

$\frac{2}{3} \div \frac{4}{1}$

$\frac{2}{3}$ (of a whole) shared equally between 4 is seen to be $\frac{2}{12}$ of the whole

$\frac{2}{3} \div \frac{4}{1}$ is the same as $\frac{2}{3} \times \frac{1}{4}$.

Algorithm

⊗ $\frac{2}{3} \times \frac{1}{4}$ Multiplication and division take place together:

$\frac{2 \times 1}{3 \times 4}$ is $\frac{2}{12}$ or $\frac{1}{6}$,

i.e. its OK to 'multiply tops' and 'multiply bottoms'.

The calculator would, *under your instruction*, carry out $(2 \times 1) \div (3 \times 4)$ or $(2 \div 3) \times (1 \div 4)$

÷ $\frac{2}{3} \div \frac{4}{1}$ Turn the second fraction upside down and ⊗.

Other suggestions:

- Lots of practice with equivalent fractions.

  $\frac{1}{2}$ is ... $\frac{37}{74}$, $\frac{550}{1100}$ ...

  $\frac{1}{3}$ is ... .

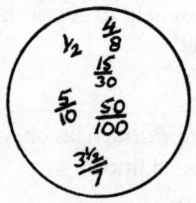

  How many kinds can they find?

  ½ $\frac{4}{8}$
  $\frac{15}{30}$
  $\frac{5}{10}$ $\frac{50}{100}$
  $\frac{3½}{7}$

- Lots of practice with converting the popular common fractions into decimal fractions.

  $1 = 1.0$, $\frac{1}{2} = 0.5$, $\frac{1}{4} = 0.25$, $\frac{3}{4} = 0.75$,

  $\frac{1}{8} = 0.125$, $\frac{1}{16} = ?$

  $\frac{1}{5} = 0.2$, $\frac{2}{5} = 0.4$, $\frac{3}{5} = 0.6$, $\frac{4}{5} = 0.8$.

What about mixing the fractions? $\frac{2½}{5} = ?$, $\frac{3½}{5} = ?$, $\frac{4½}{5} = ?$

- And give them a surprise,
  Use the calculator

  $\frac{1}{3} = 0.333333...$ $0.\dot{3}$

  $\frac{1}{7} = 0.142857\ 142857\ 142857.$

  Try $\frac{2}{3}, \frac{3}{7}, ... \frac{6}{7}$.

  Did they know that when a common fraction is expressed as a decimal fraction it either has recurring digits or is exact like $\frac{1}{4} = 0.25$?

Some questions for discussion

1. An understanding of common fractions requires practical activities such as those suggested by Piaget and Dienes. Do you agree? Is this true for all your students or only some? Do you find that practical activities can hinder the mathematical development of some students?
2. At what age should common fractions be introduced? At primary school? At secondary school? How far would you go with the development of common fractions? When would you introduce the operations of ⊕, ⊖ and ⊗, ⊘ and in what order?
3. Which would you teach first, common fractions or decimal fractions? How would you encourage your students to move easily from one form to the other?
4. What kinds of contexts can you construct for work with common and decimal fractions?

Guideline exercises

Try all these exercises with your students. Remember they are a guide from which you can construct more of your own. The attractors are designed to reflect students' ways of thinking. Can you see what they are?

Guideline exercise 1
This exercise is specifically about conversion of common fractions into decimal fractions.

$\frac{3}{4}$ written as a decimal fraction is?
a 0.34   b 0.75   c 1.3   d 3.4   e

$\frac{5}{8}$ written as a decimal fraction is?
a 5.8   b 1.6   c 0.625   d 0.58   e

$\frac{49}{100}$ written as a decimal fraction is?
a 49   b 49.100   c 0.49   d 0.049   e

$\frac{13}{32}$ correct to two decimal places is?
a 0.40   b 0.41   c 0.46   d 13.32   e

$2\frac{2}{5}$ written as a decimal fraction is?
a 2.5   b 2.4   c 2.25   d 2.2   e

Guideline exercise 2

This exercise concentrates on addition and subtraction of common fractions.

Adding $\frac{1}{4}$ and $\frac{2}{3}$ gives?
a $\frac{3}{12}$ b $\frac{2}{7}$ c $\frac{3}{7}$ d $\frac{11}{12}$ e

$\frac{5}{16}+\frac{3}{64}$ is?
a $\frac{23}{64}$ b $\frac{15}{80}$ c $\frac{8}{64}$ d $\frac{8}{80}$ e

Taking $\frac{1}{8}$ from $\frac{1}{2}$ gives?
a $\frac{1}{6}$ b $\frac{1}{4}$ c $\frac{3}{8}$ d $\frac{0}{6}$ e

$2\frac{9}{16} - 1\frac{7}{8}$ is?
a $1\frac{11}{16}$ b $1\frac{2}{8}$ c $1\frac{2}{16}$ d $\frac{11}{16}$ e

$1 - (\frac{1}{2}+\frac{1}{5})$ is?
a $\frac{5}{7}$ b $\frac{7}{10}$ c $\frac{3}{10}$ d $\frac{2}{7}$ e

Guideline exercise 3

This exercise concentrates on addition and subtraction of common fractions in practical situations.

A survey of how schoolchildren spend their pocket money, reported that they spent $\frac{1}{2}$ on sweets/snacks and $\frac{1}{3}$ on comics/magazines. The fraction left over for everything else is?
a $\frac{1}{6}$ b $\frac{2}{5}$ c $\frac{3}{5}$ d $\frac{5}{6}$ e

An educational visit is organized by a lecturer so that the Local Authority pays $\frac{1}{2}$ the bill, the college pays $\frac{1}{6}$ and the students pay the remainder. The fraction the students pay is?
a $\frac{1}{4}$ b $\frac{1}{3}$ c $\frac{2}{3}$ d $\frac{3}{4}$ e

Small numbers: common fractions 73

A company is owned by Tom, Dick and Sally. If Tom owns $\frac{1}{3}$, Dick $\frac{1}{4}$, then the fraction Sally owns is?
a $\frac{5}{7}$ b $\frac{7}{12}$ c $\frac{5}{12}$ d $\frac{2}{7}$ e

A college has $\frac{1}{5}$ of the students studying arts subjects only, $\frac{3}{8}$ studying science subjects only with the rest studying a mixture of arts and science subjects. What fraction study a mixture of arts and science subjects?
a $\frac{17}{40}$ b $\frac{4}{13}$ c $\frac{23}{40}$ d $\frac{9}{13}$ e

A quarter of a supermarket drinks profits came from alcoholic drinks, two fifths came from fruit juices and the rest came from soft drinks. What fraction of the profit came from soft drinks?
a $\frac{3}{9}$ b $\frac{7}{20}$ c $\frac{13}{20}$ d $\frac{6}{9}$ e

Guideline exercise 4

The following non-contextual tasks are concerned with multiplication and division of common fractions.

$\frac{1}{2} \times \frac{1}{4}$ is?
a $\frac{1}{6}$ b $\frac{2}{6}$ c $\frac{1}{8}$ d $\frac{2}{8}$ e

$\frac{1}{2} \div \frac{1}{4}$ is?
a 8   b 2   c $\frac{1}{2}$ d $\frac{1}{8}$ e

$\frac{1}{4} \div \frac{1}{2}$ is?
a $\frac{1}{8}$ b $\frac{1}{2}$ c 2   d 8   e

$2\frac{1}{4} \times \frac{1}{2}$ is?
a $\frac{1}{8}$ b $1\frac{1}{8}$ c $2\frac{1}{8}$ $4\frac{1}{2}$ e

$2\frac{1}{2} \times 1\frac{1}{4}$ is?
a $1\frac{1}{8}$ b 2   c $2\frac{3}{4}$ d $3\frac{1}{8}$ e

$2\frac{1}{4} \div \frac{1}{2}$ is?
a $4\frac{1}{2}$ b $2\frac{1}{2}$ c $\frac{5}{8}$ d $\frac{2}{5}$ e

$\frac{1}{2} \div 2\frac{1}{4}$ is?
a $\frac{2}{5}$ b $\frac{5}{8}$ c $2\frac{1}{2}$ d $4\frac{1}{2}$ e

$2\frac{1}{2} \div 1\frac{1}{4}$ is?
a $\frac{1}{2}$ b 2   c 3   d $3\frac{1}{8}$ e

$1\frac{1}{4} \div 2\frac{1}{2}$ is?
a $3\frac{1}{8}$ b 2   c 1   d $\frac{1}{2}$ e

CHAPTER 5

PERCENTAGE, RATIO AND PROPORTION

'And unto one he gave five talents, to another two, and to another one; to every man according to his several ability.'

(Matthew 25 v 15)

Percentage, ratio and proportion may appear to be three strange bedfellows and indeed percentage could have been appropriately placed at the end of the last chapter. Misconcepts arising with all three topics are very closely linked, however, and therefore it makes good sense to consider them together in this chapter. Difficulties with these topics illustrate the complexity of the nature of elementary mathematics. Ideas which are intuitively obvious to the mathematician are nevertheless very difficult to convey to the average non-mathematically minded person. In attempting to explain ideas we are often trapped into language which itself can generate for our students the very confusion we are trying to avoid. Students' muddled learning can be due to our muddled teaching. Nevertheless we must seek the best compromise of which we are capable when we try to combine the essence of mathematical language, its conciseness and rigour, with peoples' everyday experience.

Percentage

%?
oh no! I can't do those

TOPS trainee

Inflation %? VAT, mortgage, rates increase ...oh!...

Householder

Percentage, ratio and proportion 75

> I'm satisfied with a modest turnover % profit?... oh just 10% I buy for £10 I sell for £100

(Rich) Businessman

> 5%

Wage increases ... equality?

Percentages occur quite often in our everyday life: sales reductions, mortgage and investment rates, VAT, service charges, inflation, wage increases and so on. And the vocational contexts are many, whether it be hairdressing, nursing or even the stock market, to name but a few examples. Yet it is becoming increasingly clear that many people, whether students at school or college or whether they are adults, flinch from any computation involving percentages. A national examination involving some 8600 candidates recently set a series of percentage tasks in contexts which affect the lives of most of us. It was found that only half the candidates could work out what their new salary would be for a given pay rise, the interest they would have to pay on a bank overdraft, a 10% service charge added to a restaurant bill, how much income tax was payable, etc. The candidates were 17 years of age or older and the numbers used in these tasks were whole numbers involving quite straightforward computation. Where then do the misconcepts arise and why do they arise?

Diagnosis

Signs and symptoms

Looking

This first example highlights the results just discussed.
Lawnmower sale: which is the best buy? (£50 is the original price)

£50	£50	£50	£50
Half price	£28 Our price	40% off	£15 off

Nearly as many candidates chose '40% off' as chose half price.

Do you see the errors in these examples?

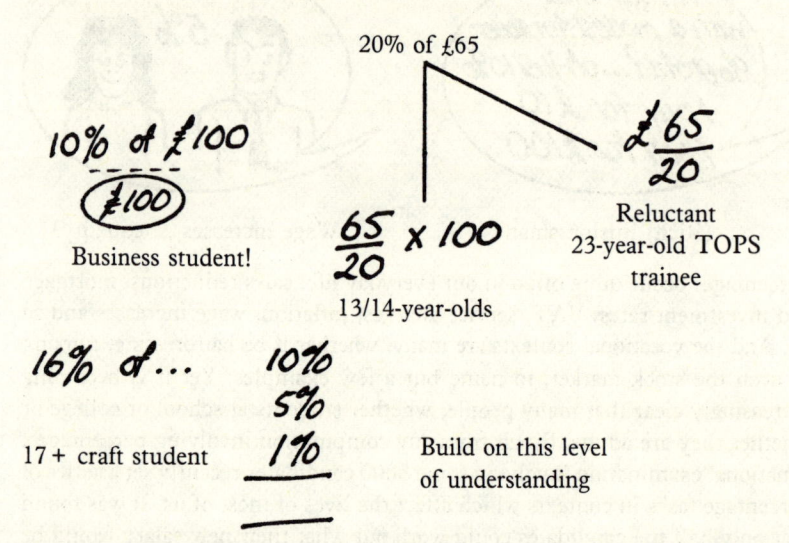

10% of £100
$\boxed{£100}$
Business student!

20% of £65
$\dfrac{65}{20} \times 100$
13/14-year-olds

$\dfrac{£65}{20}$
Reluctant 23-year-old TOPS trainee

16% of ...
17+ craft student

$\dfrac{10\%}{\dfrac{5\%}{1\%}}$
Build on this level of understanding

The following task shown in Table 5.1 was set in a context relevant to younger students.

Table 5.1
Task Everything in a sale is reduced by 20%. How much must I pay for a pair of jeans which normally cost £10?
Primary 4th-year 10+:

% choosing		% non-attempts	No. of different answers
Correct answer £8	£5		
42	12	11	36

This task shown in Table 5.2, with the normal cost price altered to £15, was given to slightly older students.

These results present rather a dismal view of future shoppers! What was judged to be a straightforward task shown in Table 5.3 was given to a variety of older students. The table shows that this task was performed satisfactorily

Percentage, ratio and proportion 77

Table 5.2
Task Everything in a sale is reduced by 20%. How much must I pay for a pair of jeans which normally cost £15?
Secondary 2nd-year 12+:

% choosing		% non-attempts	No. of different answers
Correct answer £12	£10		
26	10	20	33

by these students; there was noticeable improvement at the end of their first-year vocational course for the 17+ and there was a marked tendency for 15% to be selected as the answer especially by the older students. The recordings make clear that in many cases it was a misconcept/confusion of thought rather than carelessness which prompted this response.

The next task shown in Table 5.4 presented more difficulty for the school

Table 5.3
Task 25% of 60

Students	% choosing	
	Correct answer 15	15%
Secondary		
3rd-year 13+	65 **35**	12 **37**
4/5th-year 14/15+ (mostly 'O')	82	10
Craft/technician 17+		
Precourse	63	20
Postcourse	72	14
TOPS trainees (19–57 years)	65	30

Numbers in **bold type** refer to slower learners.

students. The older vocationally orientated students appeared to find this task if anything more to their liking than the school students and this may be the motivating effect of a relevant context. However, it must be noted that in contrast to the previous task, the students' performance did not show noticeable improvement at the end of their first year.

Table 5.4
Task 36 components out of 180 were scrapped: as a percentage this was?

Students	% choosing		
	Correct answer		
	20	5	40
Secondary			
3rd-year 13+	39 **27**	32 **30**	10 **24**
4/5th-year 14/15+	66	21	9
Craft/technician 17+			
Precourse	69		16
Postcourse	71		17

Numbers in **bold type** refer to slower learners.

Listening

A 10-year-old:

Task Everything in a sale is reduced by 20%. How much must I pay for a pair of jeans which normally cost £10?

S: I'm doing these in class. Ten pounds, reduced by twenty... Ahh... So... Em... Twenty equals... . (Mumbles)
I: What are you doing now?
S: In class we... I'm not sure... In class every one is by 10 percent.
I: How would you calculate ten percent of £10?
S: Well... we would do 10 percent of £10... that's 10 eh equals ten over one hundred equals one over ten. So ten into ten pounds

equals... that goes ten into ten... one pound... So VAT is... Ah so you take £1 from the £10. Ahh! So it would be eight pounds. (Referring to original problem)

Interview with 23-year-old TOPS trainee:

I: What is a percentage?
S: A part.
I: What is 1%?
S: A whole part.
I: What is 10% of £1?
S: Yes, that's easy ... it's 10p.
I: Why, how did you get that answer?
S: Well, ... it's 10s into 100.
I: But what does that mean?
S: One tenth of 100.
I: So, what is 1%
S: One oneth?

Task 25% of 60
Craft student 17+:

'25% of 60 is 30 ... half.'

Undergraduate 1st-year building engineer:

'Quite straightforward ... 15%.'

(Notice the slip in his response)

School student 14+:

'25% of 60 is ... 15.'

The recordings clearly showed that this task 'triggered' many students into a confidently stated automatic response, whether right or wrong. Contrast this trigger response with the 'searching around' for a solution with the next task.

Task 36 components out of a batch of 180 were scrapped: as a percentage this was?

Undergraduate 1st-year building engineers:

'Problem with these percentages... let 180 be 100% ...no ...let x be %... .'
(No clear answer emerged)
'x 100 ?... um ... just trying to remember my percentage formula... put $\frac{36}{180}$... cancel down to $\frac{1}{5}$ which equals 20%.'

(Even at this level a formula is needed as a 'prop')

Learning

Provided that students have met and carried out examples with the topic of percentage, the task **25% of 60** triggers an automatic response. A correct response of 15 suggests understanding, whereas that of 15% may suggest a knowledge of the mechanics but an inadequate concept of percentage or, of course, for the more able students it may represent a careless slip. The particular context of the task, **'36 components out of a bath of 180...**, appeared to help the further-education students. Nevertheless, the responses on the whole were not trigger ones and the recordings show more searching around by all students ...the probing of the memory for a formula.

The variety of ideas associated with percentage seems to cause so many learning difficulties that most tasks could be diagnostic. The following example, however, of two tasks which appear so similar yet contrast so strongly in performance may help to spotlight the subtlety of diagnosis.

Algorithmic	Diagnostic
25% of 60	25% reduction £60 selling price Original price?
Triggers an automatic response (70% of 17+ students got it right)	Causes a searching around (28% of 17+ students got it right)

It would appear from our studies that there are for all students tasks which trigger them reasonably confidently into a solution path which is mechanical/routine, although the students do not necessarily end up with a correct solution. These tasks appear to be amenable to improvement during a student's exposure to the normal programme of teaching. However, the diagnostic tasks, those which explore the understanding, appear to need *directed explicit* teaching of the associated concepts and skills to bring about enhanced understanding and performance.

Percentage, ratio and proportion

The diagnosis

It should perhaps not surprise us too much that people find difficulty with percentages. The misconcepts probably follow quite naturally from those associated with common fractions. Even when the tasks are related to everyday problems we are then up against another barrier, that of people's experience battling with mathematical abstraction. Mathematics seems, for many people, to be locked away in a mental compartment quite distinct from that related to everyday behaviour! For example, in the responses to tasks, such as computing 10% of a hairdressing/restaurant bill, a significant number of candidates rounded off 10% of the £ portion only, i.e. ignoring the pence. Is this what they do in practice because its easier, not worth bothering about the pence or can't compute?

The effect of context will be discussed in a later chapter but the essential diagnosis must surely be a 'lack of feeling' for the meaning of percentage. How then can we treat the inadequacy of the concepts involved?

Prescription and aftercare

Aim for our students　　　　CONCEPTS COMPETENCE and
　　　　　　　　　　　　　　CONFIDENCE
　　by ourselves being　　　　AWARE EXPLICIT and FLEXIBLE

Suggestions

Treatment needs to bring about:

- First, a thorough understanding of the fraction concept.
- Recognition and conviction that a percentage is a special kind of fraction where the total number of parts forming the whole is 100.

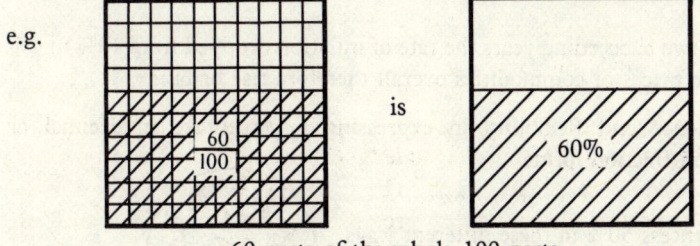

e.g.　　$\frac{60}{100}$　　is　　60%

60 parts of the whole 100 parts

Reinforcement of this is provided with the percentage sign (%) itself.
Once students have spotted the 1 and the two 0s the mystery may be on the way out!

- Give them a surprise, work out 200%, 300%, ...
- Demonstration of the usefulness and the meaning of percentage. Convincing examples are needed, show the easy comparison of quantities. (An understanding of equivalent fractions will be required.)

    e.g. A student scores $\frac{3}{20}, \frac{21}{50}, \frac{59}{100}$ in three tests. Compare these scores.

    By using equal fractions:

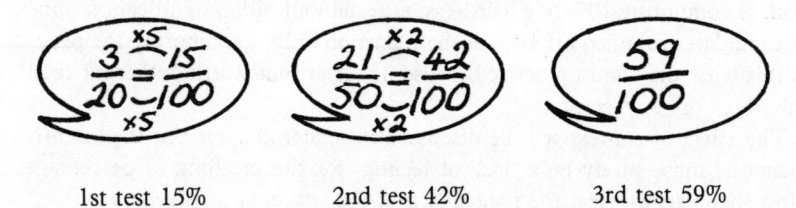

    $$\frac{3}{20} \xrightarrow{\times 5} \frac{15}{100} \qquad \frac{21}{50} \xrightarrow{\times 2} \frac{42}{100} \qquad \frac{59}{100}$$

    1st test 15%   2nd test 42%   3rd test 59%

- And further... the implication of percentage.

    e.g. A managing director of a small firm has an income of £10 000 per year whereas the foreman earns £100 per week. If each receives a pay rise of 5% who will receive the bigger increase?

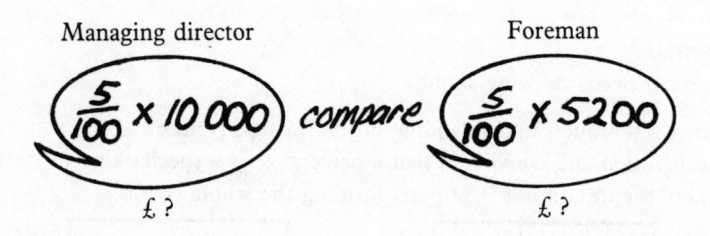

    Managing director $\qquad\qquad$ Foreman

    $\frac{5}{100} \times 10\,000$ compare $\frac{5}{100} \times 5200$

    £ ? $\qquad\qquad\qquad\qquad$ £ ?

    e.g. In two succeeding years the rate of inflation dropped from 10% to 4%. Did the prices of commodities overall therefore rise or fall?

- Confidence and flexibility in expressing a percentage in decimal or common fraction forms.

    e.g. Express 50% in these different ways with the help of equal fractions.

    $\frac{50}{100}, \frac{4}{8}, \frac{25}{50}, \frac{1}{2}, \frac{6}{12}, 50\%, 0.5, \frac{17}{34}$

How many other expressions can be found? Repeat with 20%, 25% and so on. Memorize some of these because they help in quicker problem solving:

10%	20%	25%	75%	$12\frac{1}{2}$%	$37\frac{1}{2}$%	$33\frac{1}{3}$%	$66\frac{2}{3}$%
$\frac{1}{10}$	$\frac{2}{10}\left\lvert\frac{20}{100}\right.$	$\frac{1}{4}\left\lvert\frac{25}{100}\right.$	$\frac{3}{4}\left\lvert\frac{75}{100}\right.$	$\frac{1}{8}\left\lvert\frac{12\frac{1}{2}}{100}\right.$	$\frac{3}{8}\left\lvert\frac{37\frac{1}{2}}{100}\right.$	$\frac{1}{3}$	$\frac{2}{3}$
0.1	0.2	0.25	0.75	0.125	0.375	$0.\dot{3}$	$0.\dot{6}$

Get your students to make up their own tables.
● Application to problems.
Choose contexts relevant or motivating (these are not necessarily the same) to your students. For example, for the younger/older adult, a hairdressing bill for a cut and blow-dry is £6.70. If a 10% service charge is added how much is the total bill? Problem solving needs discussion. The computation of 10% of the bill is 67p and therefore the total bill will be £7.37. This is the answer to the problem, but it will not necessarily be what a student would choose to pay in practice. In any real-life situation mathematics can supply the information, it takes us part of the way. How we use it is a matter of our own choosing. In teaching, these issues are not trivial; in discussing them the mathematics could become more meaningful for our student.

Ratio and proportion

It is not surprising that our students find difficulty in tasks involving concepts of ratio and proportion. There is some confusion of language in textbooks, examination and test questions and this may inevitably rub off in our own teaching.

Even able 'O' level pupils are muddled: as one teacher said...

'Ratio's a mystery word to them really, because they see something like 3:1 written down and one minute they're allowed to write it down as a fraction and the next minute they're not, and they get very confused about the order in which the numbers are written down quite often. If you ask them to divide something in the ratio 3:1, they're just as likely as not to do it in the other ratio 1:3. And in a question where you're asking the ratio of a part to the whole, they'll tend to give you the ratio of the two parts.'

So there is confusion between the concept of ratio and the concept of ratio as a fraction. Looking and listening at some typical tasks involving ratio may help us diagnose some of the trouble spots.

Diagnosis

Signs and symptoms

Looking at ratio and proportion

Table 5.5
Task The ratio of boys to girls in a class is 2:5. If there are 15 girls how many boys are there?

	% choosing		% non-attempts	No. of different answers
Students	Correct answer 6	10		
Primary 4th-year 10+	29	23	6	31
Secondary 2nd-year 12+	33	19	17	22

It can be seen from Table 5.5 that there was little difference in performance of the task set between the primary 10+ and secondary 12+ age groups.

The next task shown was given to older students just at the beginning of their

Table 5.6
Task 210 divided in the ratio 3:7 gives?

	% choosing			% non-attempts
Students	Correct answer 63 & 147	30 & 70	63 & 216	
Craft/technician 17+ (precourse)	41	31	13	14
Craft/technician 17+ (postcourse)	46	31	12	10

vocational course (i.e. a few months after leaving school) and again at the end of their first year. In certain respects this table echoes Table 5.5, for the younger students. The performance had somewhat improved, presumably with exposure to more mathematics: it is just possible that some of these original faint-hearted non-attempts became right answers. However, overall, the same kind of misconcepts existed at the end of the course as at the beginning.

It must be emphasized that there had been no interference with the teaching of the course, that is, no *explicit* teaching of the concepts associated with ratio had taken place over and above the usual teaching.

Listening

The listening-in process can be a valuable guide to the panic, utter frustration, and consequent blockage solution that certain mathematical tasks create.
Undergraduate building engineer solving a series of tasks involving ratio:
1st task

'...Oh,...not another of these bloody things.
... Oh, I'm just going to light a cigarette.
... yes... these... things really confuse me at moment. Leave that one... go on to next.'

2nd task

'... Oh God, another ratio one... mm... haven't a clue about that... next question.'

Our work with the older learner, even at university undergraduate level, shows only too clearly that misconcepts which arise with 'core of difficulty' tasks at school level, whether they involve, say, addition of fractions or the use of ratio can cause severe handicaps in the learning of more advanced mathematics.

More looking at proportion

The following tasks shown in Tables 5.7 and 5.8 were given to older students both at school and college.

Now compare the performance in the following related task shown in Table 5.9. All is not lost with percentage !; this appeared to be a 'trigger' task. Is it to do with the relevance of the context for these students?

Table 5.7

Task A solder is composed of tin, lead and antimony in the proportion of 25, 24 and 1 by weight. If the solder contains 400 grams of tin, the total weight of the solder in grams is?

Students	% choosing				% non-attempts
	Correct answer 800	1600	816	784	
Secondary					
3rd-year 13/14	45	12	7	9	15
4/5th-year 14/15+	54	11	15	14	6
Craft/technician 17+					
Precourse	53	5	17	17	8
Postcourse	49	7	17	15	13

Table 5.8

Task A brass contains copper and zinc in the proportion 7 parts copper and 3 parts zinc by weight. If brass has 39 grams of zinc, the total weight of the brass is?

Students	% choosing	
	Correct answer 130 g	91 g
Secondary 4/5th-year 14/15+	45	
Craft/technician 17+	33	37
Teacher trainees (primary)	45	35
Undergraduate engineers	70	26

Listening

This is the reaction of an undergraduate building technologist to the first task involving the solder:

Table 5.9
Task The percentage of copper in the brass (in previous task) is?

Students	% choosing correct answer 70
Secondary 4/5th-year 14/15+	74
Craft/technician 17+	68
Teacher trainees (all)	82
Undergraduate engineers	95

'Looks a bit more complicated... total weight is... right... get right proportion... right... 25... 400 divided by 25... No, 4,... 16 in 400... therefore take 16 off 400, makes 384. ... the sum is 800 by look of it. ... I should have noticed the question a bit better... 24 plus 1 is 25.'

This engineer had eventually seen a quicker route to solution but the fact that even he, with a good mathematics background, took some time getting there illustrates the complexity with which most students will see tasks such as these.

Learning

There is considerable evidence to show that almost any task relating to the concepts of ratio and proportion, creates problems for our learners. The major difficulty may well be that these concepts are linked in most problem solving not only together but also with the fraction concept. It is rather like trying to sort out the individuals sitting on a roller-coaster at the fun-fair! Pulling the problem apart to decide on the strategies to be adopted can create panic and confusion at the initial perception: is it ratio/fraction/proportion and if so what are these anyway?

The diagnosis

There is evidence from 'listening-in' to the recordings and from teachers that both the *word* 'ratio' and the notation ⊙ create a blockage. With this kind of reaction, treatment must be directed to creating healthy concepts of fraction, ratio and proportion.

Prescription and aftercare

Aim for students   CONCEPTS COMPETENCE and CONFIDENCE
       for ourselves   AWARE EXPLICIT and FLEXIBLE.

Clear explicit teaching of the meaning of
- Ratio
- Ratio as a fraction
- Proportion.

The acquisition of these concepts will, just as with percentage, depend on a sound understanding of *fraction*.

Suggestions

Ratio

A ratio is the comparison of two like quantities in such a way that one is a multiple of the other. The notation used for ratio is ⊙. If this definition is clearly explained by using lots of examples then few problems ought to arise with the use of ratio alone. For example, in a camera, the ratio of the aperture of the lens to its focal length may be 3:1 and that's that!

If tasks involve comparisons ratio is mostly expressed as ? : 1. Writing it in this way also defines 'scale-factor', a very useful idea in scale-drawing or in examples such as the following.

School A has 600 girls and 400 boys

Ratio of girls to boys is	600 : 400
or more concisely, ratio is	6 : 4
or	3 : 2
or often more usefully	1.5 : 1

When computing with ratio, ⊙ behaves as a ⊕ operator. For more awkward numbers the calculator can be used to express the ratio in the form ? : 1. Give plenty of practice with straight ratio tasks.

Percentage, ratio and proportion 89

School B has 800 girls and 600 boys
    Ratio of girls to boys is 800 : 600
                              or   8 : 6
                              or   4 : 3
                              or $1.\dot{3} : 1$
(cf. school A, ratio is 1.5 : 1)

The usefulness of decimal fractions is also seen here: 1.5 is easier to compare with $1.\dot{3}$ than is $\frac{3}{2}$ with $\frac{4}{3}$.

Give plenty of practice comparing ratios. Always use the calculator for awkward numbers. In problem solving, of course, we choose the ratio to fit the problem. It may be easier to use 4:3 rather than $1.\dot{3}:1$ as in some tasks with similarity, for example.

Ratio as a fraction

Ratio is probably most useful and is met most often in problem solving when it is used as a fraction. In this case the ratio must be the *comparison of a part to the whole*.

For example, in school A with 1000 students, girls/boys ratio is 3:2 (by using the simplest ratio involving whole numbers).

Ratio of girls to total is 3 : 5
Ratio of boys to total is 2 : 5
This means that three fifths, $\frac{3}{5}$, of the total number of pupils will be girls and two fifths, $\frac{2}{5}$, boys.

(Check : $\frac{3}{5}$ of 1000 is 600
              $\frac{2}{5}$ of 1000 is 400)

Emphasize, and reinforce with lots of practice, situations in which ratio can be used as a fraction.

Other examples of ratio together with its use as a fraction are of course to be found in scale drawing as in maps and model building. If the scale is 1 : 50 000, as in an ordnance survey map, then the length of a line on the map is only $\frac{1}{50\,000}$ of the true distance. In examples such as these we are actually using the concept of proportion.

Proportion

Proportion is the equivalence of pairs of ratios.
With symbols,

$\frac{a}{b} = \frac{c}{d}$ hence $ad = bc$

This of course is where 'cross multiplication', seen only as a useful 'trick' by students, comes from.

There are some helpful points to emphasize when solving tasks involving proportion. First, wherever possible, use the unknown as the first number in the ratio: it is then 'on top' in the computation. Secondly, arrange the dimensions on each side of the equivalence to be the same.

e.g. A train travelling at constant speed covers 240 miles in 3 hours. How far will it travel in 2 hours?
Keeping in mind both hints

? : 240 = 2 : 3 (ratio of distances = ratio of times)

$\frac{?}{240} = \frac{2}{3}$

$\frac{?}{80} = \frac{2}{1}$

? = 160

Distance travelled is 160 miles

e.g. Three tickets for a theatre play cost £15: what would be the cost for 12 people for the same kind of seat assuming no discount?

? : 15 = 12 : 3 (ratio of costs = ratio of number of tickets)

$\frac{?}{15} = \frac{12}{3}$

$\frac{?}{15} = \frac{4}{1}$

? = 60

Cost of 12 tickets is £60

Be flexible: this is an example of only one algorithm for solution. There are of course others which you or your students may prefer. One of these is the 'unitary' method.

e.g. In 1 hour the train will travel 80 miles
    In 2 hours the train will travel 80 × 2 miles
    ... and similarly for the cost of tickets.

Whichever way you select at first, give students considerable practice with one method at a time. Given the confidence generated with one well-tried method they may then use another and eventually be able to select according to the problem, thus acquiring the flexibility for which we all aim.

Some questions for discussion

1. Why do tasks involving percentage cause so many problems for students? Do you find it so? Is it to do with their misunderstanding of fractions? What suggestions have you for helping your students acquire clear concepts?
2. Do you find that real-life problems help your students with their understanding of percentage? What other contexts do you invent? Discuss the effect of context. Do you encourage your students to bring examples from newspapers, journals, etc., for class discussion?
3. Some confusion exists concerning the definitions of ratio and proportion. Are you clear in your mind about the use of these terms? Discuss with your colleagues how you think they should be defined.
4. Some *older* traditional textbooks have formal definitions of terms such as ratio and proportion. Do you ever look back at these? Do you consider that you could learn from such texts? Have you thought of extracting what is appropriate and fitting it into your own contexts?
5. In problem solving we often 'flip' quickly between the terms ratio, scale-factor, fraction and proportion. Is your own language clear and precise when teaching and using these ideas?
6. In the guideline exercises which follow, you will find the term 'ratio' used in the way it is usually presented in other texts. Do you think, however, that the term 'proportion' would be more accurate and appropriate?

Guideline exercises

Try all these exercises with your students. Remember they are a guide from which you can construct more of your own. The attractors are designed to reflect students' ways of thinking. Can you see what they are?

Guideline exercise 1

This exercise concentrates on finding $x\%$ of y (!)

25% of 120 is?
a 25	b 30	c 40	d 60	e

20% of 200 is?
a 100	b 40	c 20	d 10	e

15% of 17 is?
a 1.02	b 1.7	c 1.785	d 2.55	e

26% of 580 is?
a 26	b 145	c 150.8	d 290	e

0.5% of 500 is?
a 25	b 5	c 2.5	d 0.5	e

Tasks in context:
A shopkeeper reports that 25% of the delivery of eggs he received were broken. If he received a delivery of 1200 eggs the number broken was?
a 25	b 300	c 400	d 600	e

A car dealer offers a 20% discount on an old car if you pay by cash. The car is priced at £350; if you pay by cash it will cost?
a £280	b £315	c £330	d £343	e

Before VAT is added to a bill the total is £17. If 15% VAT is added to the bill the amount to be paid is?
a £17.15	b £18.02	c £18.70	d £19.55	e

If your monthly gross pay is £580 and you pay 26% of this in income tax, the amount of income tax paid is?
a £145.00	b £150.80	c £260.00	d £429.20	e

In making up a 500 litre tank of chemical 0.5% is oil. How many litres of oil are required?
a 25	b 5	c 2.5	d 0.5	e

Percentage, ratio and proportion 93

Guideline exercise 2

This exercise concentrates on finding percentages.

12 out of 40 as a percentage is?
a 3%   b 12%   c 28%   d 30%   e

$\frac{1}{3}$ as a percentage is?
a 3%   b 13%   c 30%   d $33\frac{1}{3}$%   e

0.35 written as a percentage is?
a 0.0035%   b 0.35%   c 3.5%   d 35%   e

24 out of 480 as a percentage is?
a 48%   b 24%   c 20%   d 5%   e

15 out of 75 as a percentage is?
a 60%   b 20%   c 15%   d 5%   e

Possible contexts, starting with the usual.

A test score of 12 marks out of 40 as a percentage is?
a 30%   b 28%   c 12%   d 3%   e

It was reported after a survey that one third of the home owners questioned painted the front of their house every 4 years. As a percentage this represents?
a $33\frac{1}{3}$%   b 30%   c 13%   d 3%   e

24 components out of 480 are scrapped. What percentage is scrapped?
a 5%   b 20%   c 24%   d 48%   e

In an hour and a quarter of television programmes, 15 minutes is used for commercials. What percentage of the time is spent on commercials?
a 20%   b 15%   c 12%   d 10%   e

A first-year group of university students had 10 students with 'O' level maths and 40 students with 'A' level maths. What percentage of the students as a whole had 'O' level maths?
a 5%   b 10%   c 20%   d 25%   e

Guideline exercise 3

This exercise concentrates on ratio and proportion

16 divided in the ratio 3:5 gives?
a 6 and 10   b 3 and 5   c 1 and 8   d 8 and 16   e

420 divided in the ratio 7:3 gives?
a 413 and 417
b 294 and 126
c 213 and 217
d 70 and 30
e

27 divided in the ratio 4:5 gives?
a 4 and 5   b 9 and 18   c 12 and 15   d 13 and 14   e

A brass contains copper and zinc in the proportion of 7 parts copper and 3 parts zinc by weight. If the brass contains 42 g of zinc the total weight of the brass is?
a 60 g   b 98 g   c 140 g   d 420 g   e

A business is owned by two people who agree to share the profits in the ratio 4:5. If they make a profit of £27 000 the amount each person receives is?
a £4000 and £5000
b £6750 and £5400
c £12 000 and £15 000
d £13 496 and £13 505
e

A model boat is made to a scale of 1:50. If the model is 100 mm long then the true length of the original boat in metres was?
a 1 m   b 2 m   c 5 m   d 51 m   e

Mortar can be made up by mixing sand and cement in the ratio 3:1 and then mixing with water. If there are four buckets of cement, how many buckets of sand are needed to be able to make the mortar?
a 16   b 12   c 7   d 4   e

An alloy is made up of tin, copper and nickel in the ratio 12:10:3. If the alloy contains 50 g of copper, the total weight of the alloy is?
a 5 g   b 75 g   c 125 g   d 500 g   e

Something slightly different! With a map which has a scale of 1:50 000 the distance from Egham to Hayes is 20 kilometres. If the scale on another map is 1:25 000 the distance from Egham to Hayes will be?
a 5 km   b 10 km   c 20 km   d 40 km   e

CHAPTER 6

SHAPE AND FORM: SIMILARITY

'... his Majesty's Mathematicians, having taken the Height of my Body by the Help of a Quadrant, and finding it to exceed theirs in the Proportion of Twelve to One, they concluded from the Similarity of their Bodies, that mine must contain at least 1728 of theirs... .'

(Swift *Gulliver's Travels*)

Shape and form, symmetry and pattern, play an important role in the world around us both in nature and in the man-made. The relations between length, area and volume are critical in the harmony between form, growth and movement whether it be in the animal world or in the engineering world of aircraft and ships. So the study of similar figures and their mathematical properties should provide many fascinating opportunities for context, appealing to the realities in nature and to the imaginative stories in literature such as *Alice Through the Looking Glass*, and *Gulliver's Travels*, to name but two.

The word similar has a precise *mathematical* meaning. *Similar figures are enlargements (or contractions) of the same basic shape.* This means that each

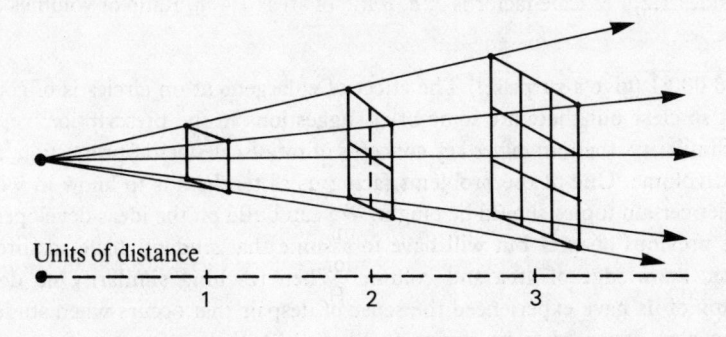

Figure 6.1 Ratio of distances 1:2, ratio of areas 1:4, and so on.

dimension of the figure is increased (or decreased) in the same ratio. The projection of light on a screen provides a very effective visual demonstration (enlist the aid of your science colleagues) (Figure 6.1).

This will also help to explain why enlargements of photographs apparently cost so much. The *punch* 'barebones' to emphasize are that:
'In an enlargement
 the area goes up by the square
 the volume goes up by the cube'.
Examples of similar figures are squares, equilateral triangles, circles and, in three dimensions, cubes, prisms and spheres (Figure 6.2). Ratio of volumes is $1:2^3$, i.e. volume scale-factor is 8. Generalize for a scale-factor of k. Choose a dramatic example of an elephant and a mouse. Assume they are of similar shape with lengths in the ratio 60:1. The volumes will be in the ratio of

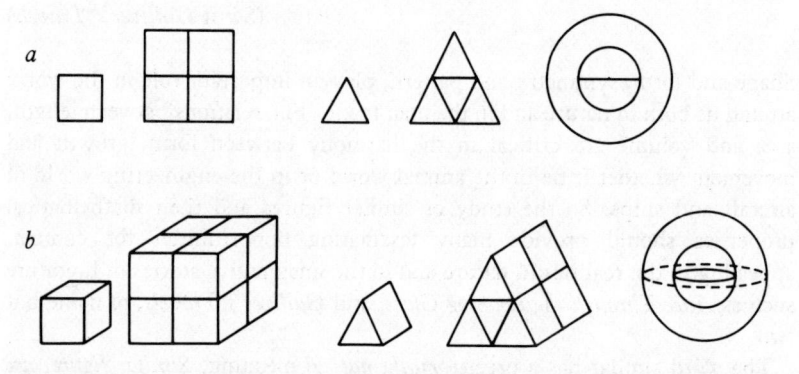

Figure 6.2 Examples of similar figures in two and three dimensions. Ratio of sides 1:2, i.e. scale-factor is 2. *a*, Ratio of areas 1:4. *b*, Ratio of volumes 1:8.

216 000:1 (give a surprise!) The effect of enlargement on circles is of course not so clear but there are some brief suggestions in the prescription.

Similarity, then, involves key concepts of ratio/scale-factor, proportion, area and volume. One of the problems facing us as teachers is to know in which order certain topics should be taught. We can build on the ideas developed in the previous chapter but will have to assume that students have acquired a basic knowledge of area and volume. When teaching similarity no doubt many of us have experienced the sense of despair that occurs when students who have appeared to be coping successfully with set after set of exercises, appear a week or two later as if they had never acquired any feeling for the

Shape and form: similarity 97

basic ideas. The concepts seem to have flown out of the window or perhaps it is rather that students never did see the wood for the trees. What misconcepts, then, arise in similarity?

Diagnosis

We have investigated many tasks in various forms appropriate to younger and older students; some of the results have been tabulated in the following discussion. *These tables need to be studied quite closely to see how apparently similar tasks can generate marked differences in performance.* And when we 'listen-in' to the solution routes this difference is emphasized quite clearly.

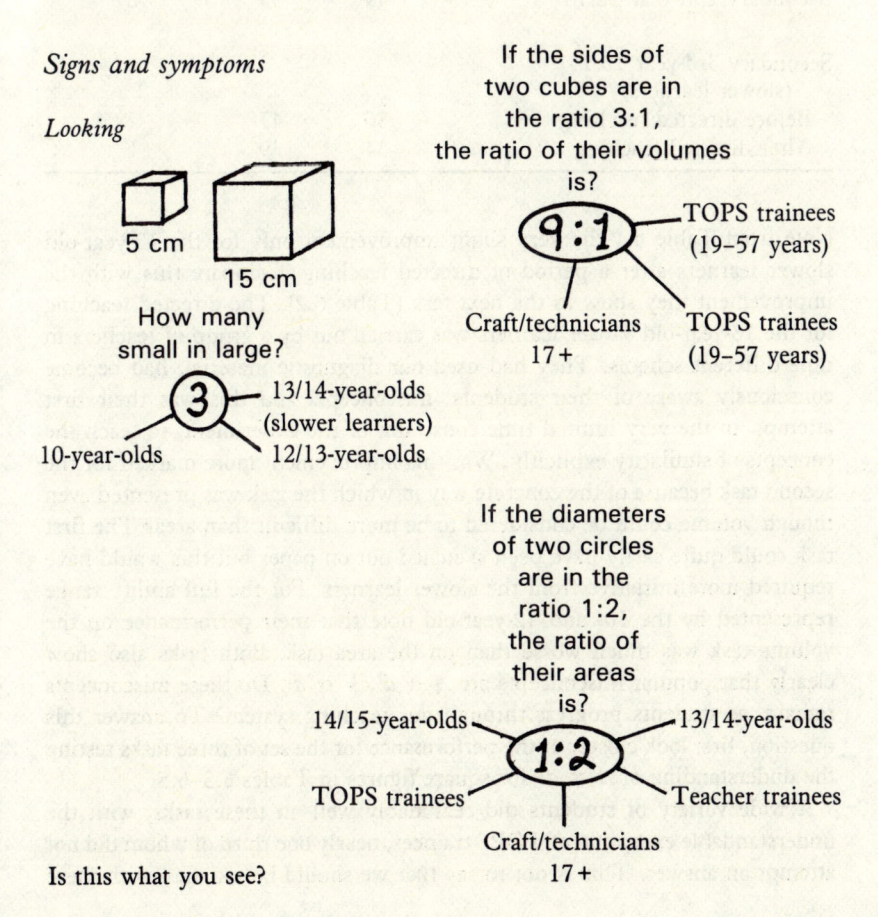

Signs and symptoms

Looking

5 cm
15 cm

How many small in large?

③ — 13/14-year-olds (slower learners)
10-year-olds   12/13-year-olds

If the sides of two cubes are in the ratio 3:1, the ratio of their volumes is?

⑨:1 — TOPS trainees (19–57 years)

Craft/technicians   TOPS trainees
17+   (19–57 years)

If the diameters of two circles are in the ratio 1:2, the ratio of their areas is?

14/15-year-olds ⟶ ①:② ⟵ 13/14-year-olds
TOPS trainees   Teacher trainees
         Craft/technicians
         17+

Is this what you see?

Table 6.1
Task How many squares with sides of length 3cm could I fit into a square with sides of length 6cm?

Students	% choosing		% non-attempts
	Correct answer 4	2 ($A \propto d^1$)	
Primary 4th-year 10+	39	37	5
Secondary 2nd-year 12/13	49	35	8
Secondary 3rd-year 13/14 (slower learners)			
Before directed teaching	30	47	2
After directed teaching	34	39	1

Note from Table 6.1 the very slight improvement only for the 13-year-old slower learners after a period of directed teaching. Compare this with the improvement they show in the next task (Table 6.2). The directed teaching for the 13-year-old slower learners was carried out by a group of teachers in nine different schools. They had used our diagnostic material, had become consciously aware of their students' misconcepts and this was their first attempt, in the very limited time constraint of the experiment, to teach the concepts of similarity explicitly. Was this improvement more marked for the second task because of the concrete way in which the task was presented even though volume could be considered to be more difficult than area? The first task could quite easily have been sketched out on paper but this would have required more initiative from the slower learners. For the full ability range represented by the 10- and 12-year-old note that their performance on the volume task was much worse than on the area task. Both tasks also show clearly that popular misconcepts are $A \propto d^1$, $V \propto d^1$. Do these misconcepts remain, as students progress through our teaching system? To answer this question, first look closely at the performance for the set of three tasks testing the understanding of $A \propto d^2$ for square figures in Tables 6.3–6.5.

A wide variety of students did reasonably well on these tasks, with the understandable exception of TOPS trainees, nearly one third of whom did not attempt an answer. This is not to say that we should be satisfied with these results; far too many students, including those on craft and technician courses, selected the response indicating the misconcept $A \propto d^1$. Compare these results with the two tasks in Tables 6.6 and 6.7.

There is a dramatic drop in the success rate for all students (except the design technologists) reflecting the difficulties with volume experienced by the younger students. The most popular error made is associated with the misconcept $V \propto d^2$ ... they have at least moved on in their thinking from $V \propto d^1$.

Tables 6.8–6.10 consist of a set of three tasks testing $A \propto d^2$ for *circles*: compare them with Tables 6.3–6.5 testing $A \propto d^2$ for *squares*. Attractors involving π^2 are of course possible with circles. We have found that many students display a certain measure of sophistication in that while they may not

Table 6.2

Task I have a collection of cubes like the one below. Each cube has sides of length 5 cm. How many of these cubes could I fit into a cube-shaped box whose sides are 15 cm long?

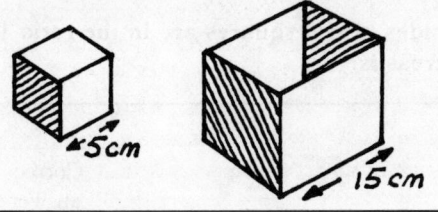

Students	% choosing			% non-attempts
	Correct answer 27 ($V \propto d^3$)	3 ($V \propto d^1$)	9 ($V \propto d^2$)	
Primary 4th-year 10+	17	28	6	5
Secondary 2nd-year 12/13	24	24	20	3
Secondary 3rd-year 13/14 (slower learners)				
Before directed teaching	18	42	18	5
After directed teaching	33	28	14	2

Table 6.3
Task If the sides of two squares are in the ratio 1:3 the ratio of their areas is?

Students	% choosing		% non-attempts
	Correct answer 1:9 ($A \propto d^2$)	1:3 ($A \propto d^1$)	
Secondary 3rd-year 13/14	51	14	16
TOPS trainees (19–57 years)	37	22	29
Undergraduate design technologists	78	12	3

Table 6.4
Task If the sides of two squares are in the ratio 1:2 the ratio of their areas is?

Students	% choosing	
	Correct answer 1:4 ($A \propto d^2$)	1:2 ($A \propto d^1$)
Craft/technician 17+ Precourse	45	28
Postcourse	51	27

understand how to find the area of a circle, nevertheless, they feel that π and the circle are so inseparable that, when it comes to area, π^2 must be involved. You will note that the success rate for this set of tasks is lower for all groups of students and that the misconcept $A \propto d^1$ is still popular although challenged by π^2. Another noteworthy point is that for those craft and technician students tested both before and after their first-year vocational course there was no improvement in the success rate (in fact slightly worse) but *more*

Table 6.5
Task If the sides of two squares are in the ratio 2:3 the ratio of their areas is?

Students	% choosing	
	Correct answer 4:9 ($A \propto d^2$)	2:3 ($A \propto d^1$)
Secondary 4/5th-year 14/15+	69	23
Craft/technician 17+	56	35
Teacher trainees (primary, secondary and inservice)	71	
Undergraduate engineers	96	3

Table 6.6
Task If the sides of two cubes are in the ratio 3:1 the ratio of their volumes is?

Students	% choosing			% non-attempts
	Correct answer 27:1 ($V \propto d^3$)	9:1 ($V \propto d^2$)	3:1 ($V \propto d^1$)	
Secondary 3rd-year 13/14	18	42	14	16
TOPS trainees (19–57 years)	20	23	2	34
Undergraduate design technologists	72	20	5	2

students attempted the task with a subsequent upsurge in the response $A \propto d^1$. It is open to debate as to whether or not there is progression or regression!

Table 6.7
Task If the sides of two cubes are in the ratio 3:2 the ratio of their volumes is?

Students	% choosing			% non-attempts
	Correct answer 27:8 ($V \propto d^3$)	9:4 ($V \propto d^2$)	3:2 ($V \propto d^1$)	
Secondary 4/5th-year 14/15+	39	28	21	7
Craft/technician 17+				
Precourse	25	36	18	12
Postcourse	22	37	28	7

Table 6.8
Task If the diameters of two circles are in the ratio 1:3 the ratio of their areas is?

Students	% choosing			% non-attempts
	Correct answer 1:9 ($A \propto d^2$)	π^2:3	1:3 ($A \propto d$)	
Secondary 3rd-year 13/14	39	23	16	16
TOPS trainees (19–57 years)	15	22	22	34

Tasks such as those discussed have been given to different school students and different college students over several years and the picture has remained reasonably consistent. This is strong evidence that the misconcepts are not only common but tenacious and require both effective diagnosis and prescription.

At this stage sympathy must be expressed for our readers having been asked to examine closely so many tables. However, looking searchingly at students' results and pausing for reflection can help us as teachers to acquire, as

Table 6.9
Task If the diameters of two circles are in the ratio 2:3 the ratio of their areas is?

Students	% choosing			% non-attempts
	Correct answer 4:9 ($A \propto d^2$)	2:3 ($A \propto d$)	$2\pi^2:3$	
Secondary 4/5th-year 14/15+	47	21	21	5
Craft/technician 17+				
Precourse	24	25	30	14
Postcourse	22	38	25	8

Table 6.10
Task If the diameters of two circles are in the ratio 1:2 the ratio of their areas is?

Students	% choosing	
	Correct answer 1:4 ($A \propto d^2$)	1:2 ($A \propto d$)
Secondary 4/5th-year 14/15+	49	
Craft/technician 17+	27	42
Trainee teachers (all)	43	
Undergraduate engineers	87	7

perhaps in no other way, a conscious awareness of our students' misunderstanding. We all only too often feel overwhelmed by data but if only

Table 6.11

Task A sphere has a volume of 3 m³. Another sphere having twice the diameter has a volume in m³ of?

Craft/technician 17+:

Students	% choosing				% non-attempts
	Correct answer 24	($V \alpha d^1$) 6	($V \alpha d^2$) 9	(3 x 3 x 3) 27	
Precourse	20	10	28	23	19
Postcourse	11	18	35	23	13

we would pause more, look and listen!

Older students were set tasks such as shown in Table 6.11. Only 11% of the craft students could do this task at the end of their course and even the university undergraduates found difficulty with such tasks as will be observed from the recordings.

Listening

Listening in to our students' solutions not only fills in the gaps in their written work but reveals ways of thinking which may surprise us.

Here are some 10-year-old fourth-year primary-school students solving the tasks discussed in Tables 6.1 and 6.2.

1st student:

Task How many squares of length 3 cm could I fit into a square with sides of length 6 cm?

S: (Reads aloud) So... four!
I: Why is that true?
S: Three... (draws square and indicating sides) this is six cm, six, six, six. Then you have three, three, three, three, three, three, three. One there, one there, one there, one there (drawing in smaller squares).
I: I see, that's four.

2nd student:

S: There are four sides in a box... a square and you times all the sides. So it would be twenty four.
I: Twenty four? Why?
S: Four times six.
I: So you have added the length of the four sides? And that's how many squares will fit in?
S: (Pause)... No... You divide that by three.
I: Why do you divide by three?
S: So that you get the answer.
I: But why did you decide on three?
S: Because the little squares are three centimetres...
The answer will be eight.

Task **I have a collection of cubes like the one below. Each cube has sides of length 5 cm. How many of these cubes could I fit into a cube-shaped box whose sides are 15 cm long?**

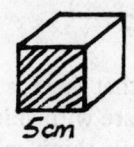

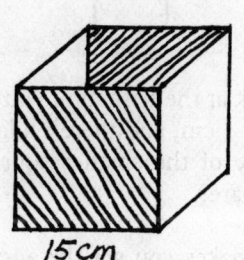

1st student:

S: (Reads problem aloud)
I: Do you understand.
S: I think so.
I: Do you understand the picture?
S: (Looks again) Oh, yes, thats 15 cm up, so its square, so thats it, and that... (indicating sides of box). Thats 3, 3, 3, so you can fit... 9 on one side. Six... nines... fifty four. I think.
I: So the answer is?
S: Fifty four.

2nd student:

S: (Reads problem aloud)... So... five centimetres... I suppose it would be again...eh... *three*... because five goes into fifteen three times.

I: So if you had a big box with sides 15 cm long and little cubes 5 cm long you could only get three of them into the box?
S: Oh... five centimetre cubes... (interviewer then checked that student understood question). So if its fifteen each... so if that's a cube (box) fifteen... every side is fifteen... then so... you are trying to fit... so you are timesing length, breadth... so fifteen, fifteen... five... twenty five... seven... plus seven is twelve...plus... One plus one is two... So two hundred and twenty five... . (He divides by twenty five) So I think you can fit nine into it. Yes. You can only fit nine into it.

Guy, 11-years-old, had just started secondary school. He was shown a picture of a smaller square and a larger square.

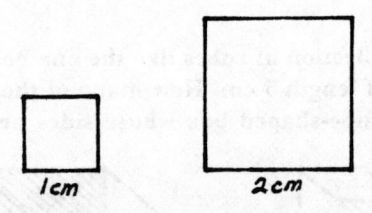

I: If you look at the first sum, you will see that there is a small square with sides 1 cm, and there's a larger square with sides 2 cm. Now, how many of the smaller squares do you think will fit into the larger square?
G: 4.
I: 4. What makes you get the answer so rapidly?
G: Well, 1 cm times 1 cm gives you the area and that's 1, and 2 times 2 is 4, so 1s into 4 is 4.
I: So you're actually calculating the area of each square to get the answer?
G: Yes.
I: Will you have a go at this question? The smaller square has a side of 3 cm and the larger square a side of 6 cm. How many of the smaller squares do you think would fit into the larger square?
G: 4.
I: Why do you say that?
G: Well, it's got all sides 3 cm, 3 times 3 cm would be 9, and 6 times 6 is 36 and 9s into 36 goes 4.

You will see that Guy calculated the areas of the squares to solve the task. Referring to this kind of work at his primary school he said:

'We used to count the little squares in the smaller one and then count them all in there, and divide how many were in there, into that.'

Sketching on paper works fine for squares but not for circles as 9-year-old Jennifer said:

'Well, we've learnt things like you can't measure areas with circles, because they leave gaps in the middle.'

By the age of 13 years many students, except possibly the slower learners, will have been introduced to ratio and tasks which are written more formally rather than in the concrete form with pictures. Neil at 13-years-old was one of the more able students and solved the following:

Task **If the sides of two cubes are in the ratio 1:3, the ratio of their volumes is?**

I: Now, how would you work that out?
N: Well, if you had a side of a cube which is 1, then the other cube would have the side of 3. Then you multiply both by 3 to get their volumes.
I: So, how would you work it out?
N: Well, it would be 1 times 3 which is 3, and 3 times 3 which is 9.
I: So what would be the ratio of their volumes?
N: Be 3:9.
I: Could you put that in a simpler way?
N: Yes, you can reduce it to 1:3 again.

So Neil was back to the original ratio of the lengths. Even older students who are university engineering undergraduates with a background qualification in 'A' level mathematics find themselves confused when faced with tasks such as these:

Task *a* **If two triangles of similar shape have sides in the ratio 2:3 the ratio of their areas is?**
　　　b **If the diameters of two circles are in the ratio 2:3 the ratio of their areas is?**
　　　c **If the diameters of two spheres are in the ratio 1:2 the ratio of their volumes is?**

Here are recordings from some of these students.

1st student:

a 'Honestly don't want to baffle myself with triangular ratios and circular ratios — apart from those I've concluded the test.'

b '...area ... $\frac{\pi d^2}{4}$
ah how do I work out this ratio? Square of ratio ... look back ... yes right ... not time... .'

c '... another one of these ... now trying to think of volume of sphere ... $\frac{\pi d^2}{4}$ ... $\frac{\pi d^2}{2}$ something like that ... 1^2 ... 1:4 πd^3 ... can't remember equation for volume of sphere ... can't recall ... leave it out.'

This student did not like ratio tasks in any form!

2nd student:

a 'Have look — draw triangle — similar shape ... area of triangle is $\frac{1}{2}h$ × base 2:3 if increase anything by $1\frac{1}{2}$ right, ... area will be $1\frac{1}{2} \times 1\frac{1}{2} \times 1\frac{1}{2}$ is $\frac{3}{4} \times 1\frac{1}{2}$ equals $\frac{3}{4}$ plus $\frac{1}{2} \times \frac{3}{4}$ equals ... $\frac{6}{8}$ plus $\frac{3}{8}$ equals $\frac{9}{8}$ equals $1\frac{1}{8}$, so it increases in fact by quite a little. Check out 1 ... 1 with area $\frac{1}{2}$. Its $1\frac{1}{2}$ ... $1\frac{1}{2}$ with area of $1\frac{1}{2} \times 1\frac{1}{2}$ which is $\frac{3}{4} \times 1\frac{1}{2}$. $\frac{1}{2}$ equals ... increase by .125 ... yeah quite interesting question! So ratio is $1:1\frac{1}{8}$ which is $\frac{9}{8}$, $1:\frac{9}{8}$ *not* $\frac{2}{3}$, not $\frac{3}{2}$, is 1 to $1\frac{1}{8}$ ... which one is it? Increase by ... 12.5% ... (he keeps going) ... nothing else for it ... $\frac{9}{8}$.'

c '...ah! here we are — sphere has volume ... oh perhaps not. Can't remember what volume of sphere is ... $\frac{\pi d^2}{4}$ $\frac{\pi d^2}{2}$ $\frac{\pi d^3}{2}$ ... .'

Typically, as with younger students, several searched for a 'formula', in this case for the area of a triangle, a circle and a sphere. They made the tasks, of course, far more complicated than they need have been. The last recording was selected because it highlights this lack of 'straight' concept even for students at this educational level. It also highlights the use of the particular multiple-choice format used. The student worked out his own solution and although it did not agree with any of those provided he had the courage of his convictions and inserted his solution in space (e).

What then can we learn from all this looking and listening?

Learning

Most tasks associated with the development of concepts relating to $A \propto d^1$, $V \propto d^3$ for similar figures generate some misconceptions. However, it has been clear from looking and listening that some tasks will more effectively diagnose than others.

To explore the understanding of $A \propto d^2$:

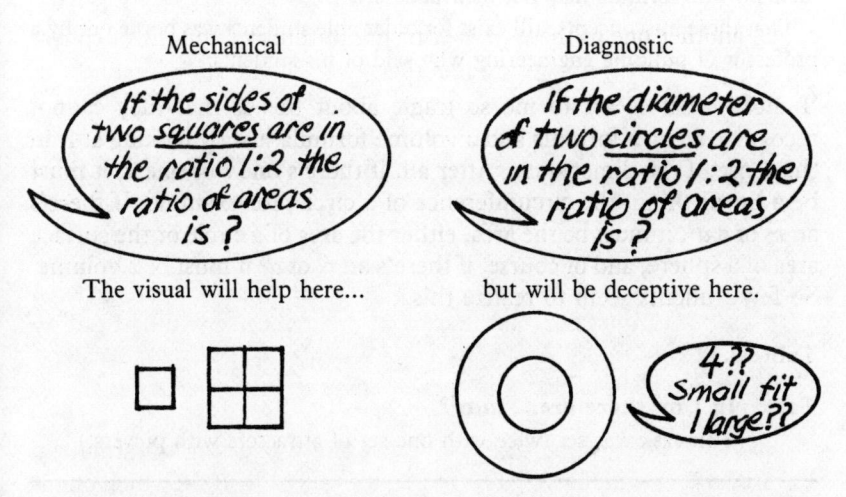

Mechanical

If the sides of two <u>squares</u> are in the ratio 1:2 the ratio of areas is ?

The visual will help here...

Diagnostic

If the diameter of two <u>circles</u> are in the ratio 1:2 the ratio of areas is ?

but will be deceptive here.

4?? Small fit 1 large??

10-year-old primary-school students had a limited and unsystematic knowledge of area and a lack of any intuitive grasp of the nature of volume in relation to a linear dimension. And these ideas of course are necessary for the later development of concepts associated with similar figures. As students progress to secondary levels, the use of ratio and other more formal mathematical language creates further problems. The common major misconcept for all students (even teachers in training), as both looking and listening have shown, is to assume that area and volume 'go up by the same ratio' as the lengths.

As one teacher of able secondary-school students said:

'It's not at all easy for them to grasp. I think they firmly believe that if two lengths are in a given ratio then everything else is in that same ratio. It's something they believe.'

But this misconcept 'wobbles' for some students as we saw: with the circle the area ratio was thought to be associated with π^2 and for volume $V \propto d^2$ is a common variation.

Students who use their earlier experience of drawing in the squares to solve tasks involving square figures correctly, nevertheless demonstrate the inadequacy of their concepts by failing to apply them to figures such as circles. Concepts have to be so strongly embedded that students' belief and conviction enable them to generalize to situations where visual back-up and memory of a formula may not be available.

That these misconcepts still exist for older able students was borne out by a professor of building engineering who said of his students:

'I mean what seems to me so tragic about this is that they cannot recognize an area formula and a volume formula just by looking at it in the sense of the dimensions. After all, if there's one r or one d, it must be a linear thing, the circumference of a circle, for example. If there's an r^2 or a d^2, it must be the area, either the area of a circle or the surface area of a sphere, and of course, if there's an r^3 or d^3 it must be a volume. So few students seem to realize this.'

Table 6.12

Task   In 1 m^2 there are... mm^2?

(This task was set twice with one set of attractors with powers.)

Students	% choosing			% choosing	
	Correct answer 10^6	10^2	10^3	Correct answer 1 000 000	1000
Secondary					
3rd-year 13/14	14	29	18		
4/5th-year 14/15+	27	29	27		
Craft/technician 17+	25	25		26	31
Teacher trainees					
All	36	28		40	
Primary	20	22	28	24	39
Engineering undergraduates	93		5	94	4

Table 6.13

Task In 1 m² there are... mm²?

Students	% choosing				% non-attempts
	Correct answer 10 000 000	100	10 000	1000	
Secondary					
3rd-year 13/14	13	28	27	24	5
4/5th-year 14/15+	29	19	30	20	2
Craft/technician					
Precourse	14	39	21	25	2
Postcourse	18	31	29	20	2
TOPS trainees (19–57 years)	11	23	27	12	20

In real life, of course, these relations are important and determine, for example, estimations of fluid flow, volume of concrete mix required, and are used in more general problems of engineering structures.

Further signs and symptoms: units of measure

In the context of school, college or work the concepts associated with similarity also spill over into the work with units of measure such as the metric system. Tables 6.12 and 6.13 illustrate this.

Looking

You will observe from Table 6.12 the popularity of the responses associated with the misconcept $A \propto d^1$. In particular the responses of the teacher trainees reinforce the strong indication of the 'learning – teaching – learning' cycle associated with the core of difficulty. The next task (Table 6.13) was found even more difficult.

112 Diagnosis and prescription in the classroom

Listening

This is an engineering undergraduate:

'... 10 000 by 10 000 ... 10 000 is ... uh ... in powers is 10^4 ... which is $10^4 \times 10^4$. Can't see an answer like this at all so will substitute my own answer which is 10^8 ... will have a think as to best one ... yes ... I'll say 10^8.'

This student as you see did not solve the task correctly. Younger or less able students find it very difficult.

One physics teacher speaking of her work with able secondary-school students said:

'When calculating area, for example, one may want to calculate in square metres or in square centimetres. Now, the pupils know there are 100 centimetres in a metre, but they find it very difficult to see that there are 100 times 100 square centimetres in a square metre. The problem also arises, of course, with cubic metres, where you have 100 times 100 times 100 and where we are then dealing with a million, and at that stage with the 11-year-old class, one seems to lose half the class because we're into large numbers.'

and one primary-school teacher said:

'It's really a matter of getting a firm understanding of the units that's the problem. And when they've been spending such a long time measuring things in one dimension, centimetres and metres, they find going into area and volume much harder.'

What treatment then could be effective for dealing with existing misconcepts or for preventing them arising?

Prescription and aftercare

Aim for students   CONCEPTS COMPETENCE and CONFIDENCE
       for ourselves   AWARE EXPLICIT and FLEXIBLE.

Teachers' comments

Helping our students to acquire the fundamental concepts of similarity is important. As a mathematics professor (Professor Zeeman in *Mathematics with Meaning*) said:

'It's very important to get the notion across of how the area goes up by the square and the volume goes up by the cube of similar shapes.'

These are the basic ideas and we have to help our students to extract them from the background 'noise' of information in so many tasks we set them. Clear thinking is needed about concepts of area, volume, ratio and proportion and we cannot begin too early as two primary-school teachers point out.

First teacher:

'Scale is a topic that has got to be done almost entirely practically, certainly when the children are young. We have children in the school who are only 8-years-old and recently made a lovely scale-model of the school, one centimetre equal to one metre. The scale-model is at present in the entrance hall of our school and it's creating an enormous amount of interest among all the children who come up to it.'

Second teacher:

'In our first year, one of the first tasks involving number and volume we get our children to undertake, involves the use of isometric paper which we find extremely useful. We get the children to initially start with cubes, and we talk about Oxo cubes and sugar cubes, and we get the children to draw on isometric paper 1 cm cubes, 2 cm cubes, 3 cm cubes and so on up to 10 cm cubes... .'

Length, area and volume all involve *numbers*. Even large ones create difficulty as our secondary-school teacher emphasized and therefore lots of early practice is needed seeing how numbers grow ... in particular when they are squared and cubed.

And of course, similarity also means figures becoming smaller, and dramatically so as small numbers are squared or cubed. We have already discussed the treatment for large and small numbers in Chapters 2–4 and the concepts discussed there will lay the foundation for the more general work on similarity.

To help our students acquire clear concepts related to similar figures our own language needs to be *mathematical*, concise, clear and explicit. We should not flinch from using 'similarity' and 'enlargement' as proper mathematical terms but we should relate them to reality and imagination.

Suggestions

The concrete approach

Both visual, and handling of materials, e.g. drawing and making models; Dienes' apparatus is helpful for work with rectilinear figures such as squares, triangles, cubes, prisms, etc. Circles and spheres prove more difficult; the visual can be deceptive as we have seen. Making, cutting out, weighing and volume displacement can be helpful.

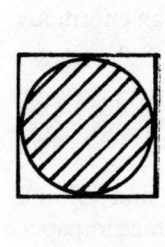

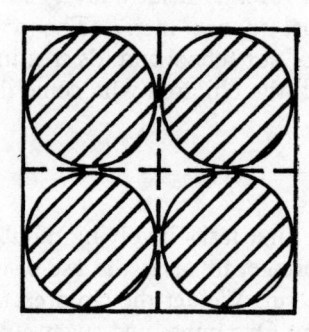

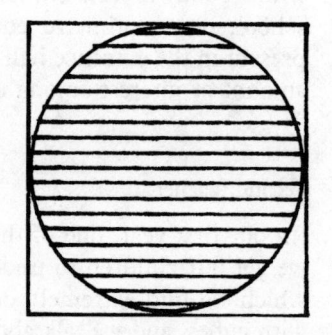

The smaller and larger circles could be cut out and weighed, even graph paper seems to be alright. You will need a ruler, paper-clips, cotton and Blu-tack.

Spheres suggest water displacement or weighing. Some bright ideas are needed here and perhaps liaison with our craft and design technology colleagues... maths across the curriculum!

Real-life applications

- Photography and projection. This is where the mathematical meaning of enlargement can really be demonstrated: doubling the distance of a screen from the light enlarges the picture 4 times and so on.
- Scale drawings, maps and model building.

Appeal to the imagination:

Give surprises as in these stories retold by Professor Zeeman:

'If you made a giant 10 times as high as a man, then he'd break his thigh bone every time he took a step, because his weight goes up by the cube and the cross-section of his thigh bone only goes up by the square, and so he'd put 10 times as much weight per unit area on his poor old thigh bone... .'

And then, of course, there's another famous example of heat in an animal:

'dinosaurs had a huge volume and yet they could only heat up or cool off by the surface area. And so, if there was a sudden climactic change and they got cold, they could never heat themselves up again and that may have been one of the reasons why dinosaurs died out.'

Discuss the practicalities of similarity with another example:

'if you wanted to fill a swimming pool with a bucket, you would take lots of journeys with the bucket and so you might think, well, why don't I use a dustbin, because a dustbin is twice as high as a bucket. But you try lifting a dustbin full of water and you'd soon realize why you have to fill it with a bucket.'

Other resources

These are available in the form of BBC TV programmes such as *Maths Topics: Geometry (Similarity and Congruence)*. Attractive posters also help. Presenting work in these ways can be dramatic, clear and form the basis for class discussion.

'Pulling it all together'

This is a very important function of our teaching. Cement the fundamental ('straight') concepts:
If the linear scale-factor is 2, the area scale-factor is 4 and the volume scale-factor is 8 (particular areas or specific volumes do not have to be calculated). And with older pupils we can extend the generalization even further. For any number that's k, if the length scale-factor is k, then the area scale-factor is k^2 and the volume scale-factor is k^3.
And the rewards for all our efforts?

School student:

'If the diameters of two circles are in the ratio 2:3 the ratio of their areas is 4:9 because area is proportional to diameter squared.'

University undergraduate:

'Cubes with sides ratio 1:2 ... volumes ratio? ... volume going to be 8 times as much, so ratio must be 1:8.'

'Similarity ... its enlarged our vision'

Some questions for discussion

1. Do you find that your students often 'do not see the wood for the trees' in work with similar figures? How in your teaching do you aim to develop the *essential* concepts and skills of similarity?
2. What do you see as the role of practical work in establishing the meaning of similarity? What suggestions for practical activities can you make?
3. Do you relate your teaching of shape and form to nature and to the man-made environment?
4. Do you relate the concepts associated with similarity to other parts of the curriculum, e.g. geography, home economics, design and technology?
5. Have you any suggestions for practical apparatus which could be used to develop the similarity concepts for circles and spheres?
6. One general aim of teaching maths should be 'to develop an awareness of the uses of mathematics in the world beyond the classroom' (*Mathematics 5–11*). Do you agree? Have you any suggestions?

Guideline exercises

Try all these exercises with your students. Remember they are a guide from which you can construct more of your own. The attractors are designed to reflect students' ways of thinking. Can you see what they are?

Shape and form: similarity 117

Guideline exercise 1

This exercise concentrates on area concepts.

The number of squares with sides of length 4 cm that would fit into a square with side length 16 cm is?
a 4　　　b 8　　　c 12　　　d 16　　　e

If the sides of two squares are in the ratio 1:4 the ratio of their areas is?
a 1:4　　　b 1:8　　　c 2:8　　　d 1:16　　　e

If the sides of two squares are in the ratio 2:5 the ratio of their areas is?
a 2:5　　　b 4:10　　　c 4:25　　　d 4:49　　　e

If two triangles of similar shape have sides in the ratio 3:2 the ratio of their areas is?
a 9:4　　　b 5:2　　　c 6:4　　　d 3:2　　　e

If the diameters of two circles are in the ratio 1:2 the ratio of their areas is?
a 1:4　　　b 2:4　　　c π^2:2　　　d 1:2　　　e

A square has an area of 1 m. Another square having sides three times as long has an area in m^2 of?
a $\frac{1}{3}$　　　b 1　　　c 3　　　d 9　　　e

Guideline exercise 2

This exercise concentrates on volume concepts.

The number of cubes with a side length 2 cm that would fit into a cube-shaped box whose sides are 6 cm long is?
a 27　　　b 12　　　c 9　　　d 3　　　e

If the sides of two cubes are in the ratio 1:2, the ratio of their volumes is?
a 1:2　　　b 1:4　　　c 1:6　　　d 1:8　　　e

A cube has a volume of 3 m^3. Another cube having sides twice as long has a volume in m^3 of?
a 6　　　b 9　　　c 24　　　d 27　　　e

Two cube-shaped boxes P and Q are measured. The side lengths of box P are twice the length of each side of box Q. The volume of P is?
a twice the volume of Q
b four times the volume of Q
c six times the volume of Q
d eight times the volume of Q
e

A and B are two spheres where the diameter of sphere B is known to be three times the diameter of sphere A. The volume of sphere B compared with the volume of sphere A means that B takes up?

a three times the volume of A
b six times the volume of A
c nine times the volume of A
d twenty seven times the volume of A
e

Guideline exercise 3

This exercise concentrates on units of measure.
The number of millimetres in 1 metre is?
a 10   b 100   c 1000   d 10 000   e

The number of square millimetres in 1 m^2 is?
a 10   b 100   c 1000   d 1 000 000   e

The number of square millimetres in 5 m^2 is?
a 5 000 000   b 25 000   c 5000   d 2500   e

The number of cubic millimetres in 1 m^3 is?
a 10^3 b 10^6 c 10^9 d 10^{12} e

The number of square feet in a square yard (three feet in one yard) is?
a 3   b 6   c 9   d 12   e

CHAPTER 7

THE CIRCLE AND ITS MEASUREMENT

'He made the sea of cast metal, circular in shape, measuring ten cubits from rim to rim... . It took a line of thirty cubits to measure round it.'
(*IInd Chronicles* book 4 v 2)

We are surrounded by circles, whether in nature or man-made objects, from planets to the atom, from the giant telescope to the camera lens. The circle is a shape we all enjoy; it is full of mystery having no beginning and no ending. It is smooth, continuous and yet functional; where would we be without the wheel or the mini-roundabout?

Yet measurement of the circle, in particular its circumference and area, seems beset with difficulties for students of all kinds. Misconcepts arise of course in varying degrees for different students with the study of perimeter, area and volume. Many of the tasks we set students, however, can be solved by them apparently successfully when related to squares, rectangles, cubes and cuboids. Our students can use a formula with which they've become 'rote-familiar' and they may even score full marks in an exercise. But if their understanding is not explored misconcepts are concealed and trouble looms ahead.

Our students cannot fool themselves or us as readily when it comes to measurement of the circle. Tasks associated with its circumference and area expose the learner's understanding of the concepts. As 9-year-old Jennifer implied (Chapter 6), circles do not tessellate. So whereas drawing squares, for example, can provide visual reinforcement of area, circles are deceptive when we try to compare them in this way. After all, the Greeks tried in vain to 'square the circle' and indeed the term 'circle squarer' came to mean someone who attempted the impossible.

There is also the additional mystery of pi (π)! Many older students have hazy recollections through the mists of time ... 'it's something to do with

Pythagoras.' Almost all are convinced of its association with the circle but is it a measure and when it comes to area is it 'π^2'? What then are the misconcepts which can arise?

Diagnosis

Signs and symptoms

Looking

We probably find that most of our students can become quite proficient with the use of 'length × breadth' when calculating the areas of squares and rectangles. But do they really understand the meaning of area? They *seem* to understand perimeter ... but do they? Altering the units of measure can help to expose their understanding.

Is this what you would see?

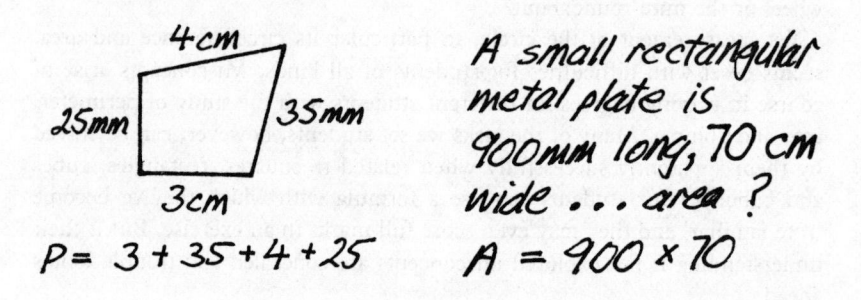

[Handwritten: quadrilateral with sides 4cm (top), 35mm (right), 3cm (bottom), 25mm (left)]

$P = 3 + 35 + 4 + 25$

A small rectangular metal plate is 900mm long, 70 cm wide ... area?

$A = 900 \times 70$

These imply instrumental understanding rather than relational understanding... our students have rules but no reasons.

The following task (Table 7.1) was given to younger students. It appears to have caused considerable difficulty: finding the length of a given square demands more than a rote calculation of area.

The concept of area needs to be more firmly established to deal confidently with such a task. The effect of age and more exposure to teaching is seen clearly from the results of a similar task given to older students.

Table 7.1
Task If the area of a square is 25 m² what is the length of each side?

Students	% choosing Correct answer		% non-attempts	No. of different answers
	5 m	25 m		
Primary 4th-year 10+	32	6	20	65
Secondary 2nd-year 12+	15	5	29	42

Task The side of a square of area 400 mm² is?

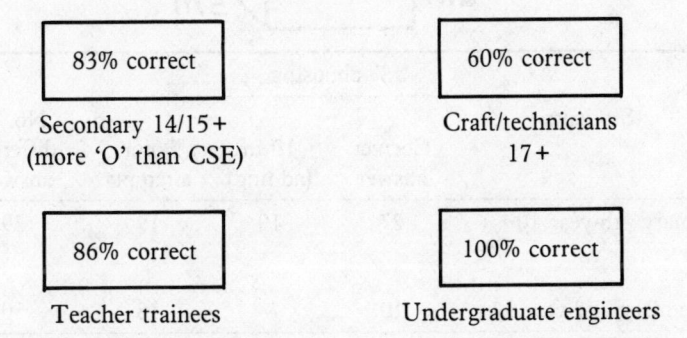

83% correct	60% correct
Secondary 14/15+ (more 'O' than CSE)	Craft/technicians 17+

86% correct	100% correct
Teacher trainees	Undergraduate engineers

Observe how the successful performance levels drop when a task, which is not too dissimilar, nevertheless demands rather more inference.

Task A rectangle 80 mm by 20 mm has the same area as a square of side?

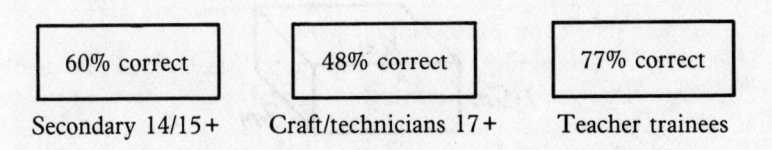

60% correct	48% correct	77% correct
Secondary 14/15+	Craft/technicians 17+	Teacher trainees

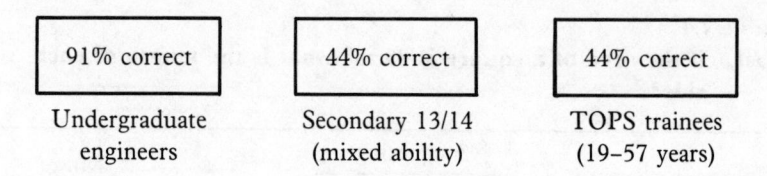

122 Diagnosis and prescription in the classroom

91% correct	44% correct	44% correct
Undergraduate engineers	Secondary 13/14 (mixed ability)	TOPS trainees (19–57 years)

If we look at tasks involving volume of a cuboid a similar picture is revealed (Table 7.2). It is clear that the concept of volume is not yet established.

Table 7.2
Task Find the volume of the shape in the figure below

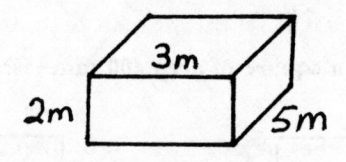

	% choosing			
Students	Correct answer	10 m (adding)	% non-attempts	No. of different answers
Primary 4th-year 10+	27	19	19	39
Secondary 2nd-year 12+ [a]	10	—	18	50

[a] For these students the task had been put into the context of a tank containing water.

This is what happened with the older students:

Task The volume of the solid shown is?

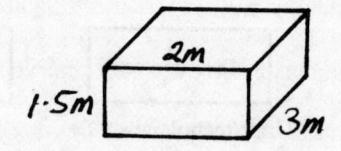

The circle and its measurement 123

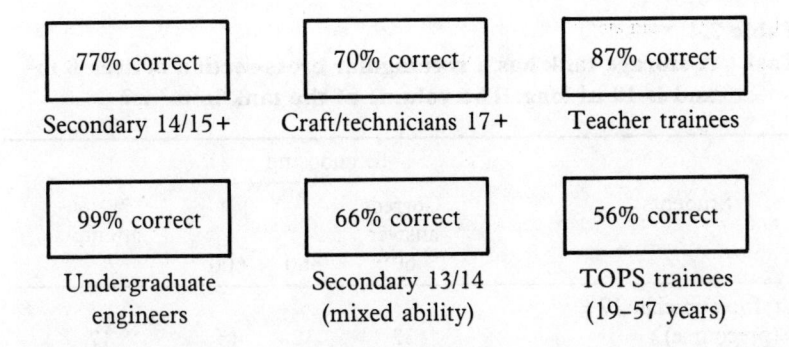

77% correct	70% correct	87% correct
Secondary 14/15+	Craft/technicians 17+	Teacher trainees

99% correct	66% correct	56% correct
Undergraduate engineers	Secondary 13/14 (mixed ability)	TOPS trainees (19–57 years)

This was a straightforward 'mechanical' task for these students.

When the tasks demand rather more inference this is what happens, as shown in Table 7.3:

Table 7.3
Task How many cubes of side 5 cm fit into a cube of side 15 cm?

Students	% choosing			% non-attempts
	Correct answer 27	3 ($V \propto d$)	9 ($V \propto d^2$)	
Primary 4th-year 10+	17	28	6	5
Secondary 2nd-year 12+	24	24	20	3
Secondary 3rd-year 13+ (slower learners)				
Preteaching	18	42	18	5
Post-teaching	33	28	14	2

- The most popular incorrect response chosen was 3, i.e. $\frac{15}{5}$.
- The secondary third-year slower learners had received *explicit* teaching on some of the core of difficulty tasks. There was a significant improvement as you will observe.
- The concept of volume is not yet developed for younger students.

The results of another task set to older students are shown in Table 7.4.

Table 7.4

Task A storage tank has a rectangular cross-section of area 6 m² and is 10 m long. The volume of the tank in m³ is?

Students	% choosing			% non-attempts
	Correct answer 60	360	600	
Craft/technician 17+ (precourse)	32	32	15	17
Craft/technician 17+ (postcourse)	41	31	17	9

Table 7.5 shows the results of 13-year-old slower learners being set a task on circumference of a circle. We expect older students to know the formula ... but do they? (Table 7.6).

Table 7.5

Task If π is $\frac{22}{7}$ what is the circumference of a circle of diameter 14 cm?

Secondary 3rd-year 13/14 (slower learners):

% choosing			% non-attempts
Correct answer 44 cm	7 cm	14 cm	
27	22	21	20

Considering most students' desire to have a formula for their problem solving, it is clear that the formula for the circumference of a circle is not so popular that it stands out above the other attractors. What is strongly indicated is the *confusion* that exists in students' minds with the circumference and area of a circle. So now let's look at tasks to do with the area of a circle. Do students know the formula for the area? (see Table 7.7).

Table 7.6
Task The circumference of a circle diameter d is?

Students	% choosing				% non-attempts
	Correct answer πd	πd^2	$2\pi d$	$\pi^2 d$	
Secondary 3rd-year 13/14	28	17	18	15	17
Secondary 4/5th-year 14/15+	45	12	25	13	5
Craft/technician 17+					
Precourse	42	20	16	14	10
Postcourse	50	20	20	7	3
TOPS trainees	25	14	10	11	30

Table 7.7
Task The area of a circle of diameter d is?

Students	% choosing				% non-attempts
	Correct answer $\frac{\pi d^2}{4}$	πd	$2\pi d$	$\pi^2 d$	
Craft/technician 17+:					
Precourse	19	33	21	21	6
Postcourse	19	39	23	16	3

Tasks associated with the circumference and area of a circle form part of the core of difficulty. Exposure to a normal programme of teaching appears to have little effect on improvement other than to encourage more students to 'have a go'. *Explicit* teaching appears to be essential for the formation of adequate concepts. Not even the motivating effect of maths in relevant contexts seems to be sufficient. Craft and technician students as part of their work need to substitute in and use formulae associated with the circle. Fabrication students, for example, need to know that the length of a piece of sheet metal becomes the circumference of a pipe.

Let's take a look at the performance of a wider range of students. Table 7.8 gives the students' choices; study these carefully. Note the popularity of the choices especially πd.

Table 7.8
Task If π is 3.142 the area in mm^2 of a circle of diameter 20 mm is?

Students	% choosing				% non-attempts
	Correct answer 314.2	62.84 (πd)	(3.142)$^2 \otimes 20$ ($\pi^2 d$)	31.42 (πr)	
Secondary					
13/14 (mixed)	10	30	20	14	4
14/15+ (mixed)	21	29	18	18	14
14/15+ ('O')	51	20	16	9	3
Craft 17+					
Precourse	20	29	20	24	7
Postcourse	19	18	13	29	21
Technicians 17+					
Precourse	31	39	18	7	5
Postcourse	27	35	17	10	11
TOPS trainees (19–57 years)	18	17	14	8	36

In the following task the value of π is not given so the computational part is virtually removed... but do they know the formula?

Task **The diameter of the base of a cone is 30 mm. The area of the base is?**

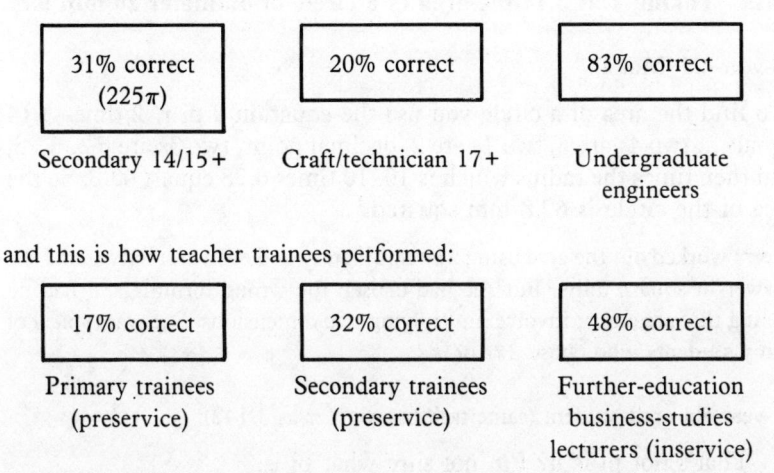

31% correct (225π)	20% correct	83% correct
Secondary 14/15+	Craft/technician 17+	Undergraduate engineers

and this is how teacher trainees performed:

17% correct	32% correct	48% correct
Primary trainees (preservice)	Secondary trainees (preservice)	Further-education business-studies lecturers (inservice)

Popular wrong responses were 30π ($d\pi$) and $15\pi^2$ ($r\pi^2$). Perhaps we should pause for reflection here... these results give some indication of the problems we face. Today's student is tomorrow's teacher; today's teacher was yesterday's student. Is it possible that one of our problems is that there exists a learning–teaching–learning cycle? The core of difficulty tasks help to expose the understanding of all of us, teachers as well as students. Even specialist mathematics teachers in training gave evidence of 'wobble' with the tasks: some lack of assurance in their routes to solution as their recordings showed.

It's true that the studies with teacher trainees were carried out before the 'O' level requirement but we have tested many undergraduates since. The two years between an 'O' level qualification and entrance to a teacher-training course is a long time in the life of a student and we have little grounds for optimism unless the time given to maths in the course is substantially increased.

What can we learn from students' wrong responses? Is it that students' thinking in fact is quite mathematically intelligent ... after all, πd and $r\pi^2$ indicate some notion that area involves squaring. But why is πd so attractive ... does it recall earlier happy memories of tying string round circular objects? It is the 'listening in' which cues us more directly into students' thinking and which provides so many of the answers to our questions, so now let's do just this.

Listening

Task Taking π as 3.14 the area of a circle of diameter 20 mm is?

13-year-old (Moira):

'To find the area of a circle you use the equation 2 pi *r*. 2 times 3.14 equals ... two 4s are 8, two 1s are 2, decimal point, two 3s are 6 ... 6.28, and then times the radius which is 10. 10 times 6.28 equals 62.8. So the area of the circle is 62.8 mm squared.'

Moira worked out the area using $2\pi r$ confidently and even correctly stated the answer in square units. But she had chosen the wrong formula and had no feeling that area must involve multiplying two dimensions. She was typical of many students who chose $2\pi r$ or πd.

17-year-old craft student (same task but with π as 3.142):

S: That's not pi is it? I'm not sure what pi is.
I: Have you met it at work but forgotten it from school?
S: Yes. (He points to mm^2 in the correct response)
I: You know what that is?
S: Yes ... sq mm ... answer is 314.2 mm^2.
I: Why?
S: It looks the same as pi but decimal points have been moved four places ... because have to multiply ... square ... square root ... taking the area is ... square mms ... it all just come to me — I thought that's it ... can't really explain it — it comes to me. But not sure why.

This student was a little muddled in his explanation but he was getting there. He was searching his memory and shades of past teaching and learning were bearing fruit. He 'felt' that the choice he had made *looked* right ... rather like a young student musician might detect the correct pitch of a note.
His thinking through this task was typical of the way he tackled others, a most enjoyable interview!

Undergraduates (building technologists):
First student:

'Area ... (pause) ... 3.142 times ... (pause) ... 314.2.'

Second student:

'Area is ... πr^2 ... therefore 3.142 × 2 which is 6.284 ... times 10 ... 62.84.'

This student stated the correct formula but then proceeded to work out $2\pi r$! (Did he remember tying string around cans?)

Third student:

'... oh that's easy. Area of circle... πr^2 ... diameter ... radius is 10 ... 100 times π ... 314.2.'

and this same student, on the task of selecting the correct formula for the area of a circle:

'... oh ... God ... $\pi d/2$ ... area is πr^2 and r equals $d/2$ therefore $\frac{\pi d^2}{4}$'.

He was fairly confident but the first two students were more typical of a certain amount of unease with these circle tasks.

Teacher trainee (undergraduate mathematics specialist):

'Area ... πr^2 ... can't see this here ... oh ... r is $d/2$ ... r^2 is ... um ... oh ... $\frac{\pi d^2}{4}$ ... oh dear ... I should have got that straight away.'

This student was taking an honours course and about to take her degree examination. Her hesitancy over $\frac{\pi d^2}{4}$ highlights the approach of the mathematician versus the engineer. School students and older mathematics students all talk in terms of 'r' the abstraction, whereas the vocational further-education students and the engineers all talk in terms of 'd', something they can handle. Sit a group of mathematicians and engineers around a table, try it out and see what happens! This is a story told by a professor of building technology:

'Oh, you know the story, I expect, about the difference between the mathematician and the engineer. The mathematician says that the area of a circle is πr^2 and the engineer says it's πd^2 upon 4. I remember I had a student once and I said to him, you know, "What is the difference between these two formulae?" And he said: "Oh, well, πd^2 over 4 is the area of a circle with a safety factor!'

But what about the mysterious 'pi', the 'circle number' which so intrigued the Greeks? One primary-school teacher said this:

'Children find this extremely hard. The Greek letter and the usage of the expression pi is a very difficult concept for them. They say: "Why can't you call it three and a bit if it's a number? Why do we have to call it pi?" And they're continually confused by this one word which is a number.'

And its not only children who are mystified. Older students of all kinds think that π ranges from something to do with Pythagoras ... to yes ... definitely to do with the circle. Some think that its a dimension so that πd really does imply an area formula.

11-year-old Martin was one of those students who really do appreciate that π is a number and a special one at that:

I: Can you tell me a little bit more about pi. Now you've written it down as 22 over 7. Is it exactly 22 over 7, do you think?

M: No, it isn't because one seventh can't be put into a decimal fraction. So it's just an in...inf... it's a number that goes on for ever.

Learning

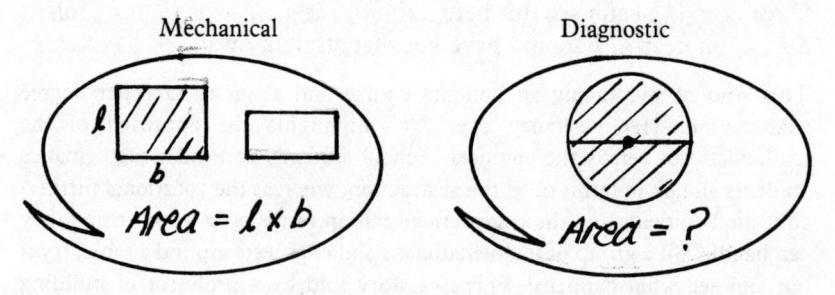

Mechanical   Diagnostic

Area = $l \times b$ Area = ?

Length × breadth OK but π ...!

Mathematics really does feed on mathematics and herein of course lies some of our problems in teaching and learning. During our interviews with students we found, apparently by chance, but probably more through the intuition gained by experience, two other stumbling blocks which help to generate the misconcepts with the area of the circle. The first one arose in algebra when we were discussing with some students the value of x^2y given x and y. Quite frequently they worked out x^2 as $2x$. Something 'clicked' in our own thinking and we led the students to talk about why they chose πd as the area of a circle.

'Well ... d is twice r...($2r$) ... and $2r$ is same as r squared (r^2) ... so area is πd.'

This is quite logical if you believe that x^2 is $2x$. Indeed one teacher told of her own experience at school and how it affected her own teaching:

'When I was at school, I was taught $2\pi r$ for the circumference of a circle, but in my first year of teaching I soon found that this became very easily confused with the formula for the area of a circle πr squared and so I eliminated $2\pi r$ from my vocabulary and we've stuck to pi times the diameter ever since, πd.'

The other stumbling block is related to commutativity. Some 'O' level students did not appreciate that πd which is $\pi 2r$ was the same as $2\pi r$. One 13-year-old girl had her eyes opened to commutativity and never looked back: she is now an enthusiastic and competent young maths teacher!

The problems with the circle are concerned with shape, symbols and formulae and these embrace several aspects of mathematical thinking. The confusion with πr^2, πd, etc., exposes a lack of feeling for symbols and their relations which follows from a lack of feeling for number. Appreciation of the language of mathematics is needed. For example, students should understand that π is a number.

The diagnosis

- Confusion of squaring and doubling, i.e. $2r \neq r^2$.
- The mystery of π ... it is actually a number.
- Misunderstanding commutativity.
- Confusion with concepts of circumference and area.
- Non-appreciation of circumference expressed as a linear dimension and area as a product of two linear dimensions.

How best can we help our students to a better appreciation of the circle and its measurement?

Prescription and aftercare

Aim for students   CONCEPTS COMPETENCE and CONFIDENCE
       for ourselves   AWARE EXPLICIT and FLEXIBLE

Treatment of course, should start as early as possible!

Suggestions

- Help students to be aware of shapes in the world around: hexagons and circles, of rings of tree trunk, architecture, domestic things, etc.
- Work with perimeter and area of many shapes, regular and irregular. Build up the concept of perimeter as the distance around a shape. (Introduce the circle and 'pi' as p in perimeter from the old Greek world periphereia.) Build up the concept of area as the amount of surface enclosed by the perimeter.
- These concepts can be developed through practical activities: string around objects of all shapes for perimeter; counting squares in traced outline of objects for area. Cut up shapes and fit them together in different ways, shapes with same areas can have different perimeters. Get students used to handling a pair of compasses and explore the geometry of the circle. Emphasize the relation between diameter and radius.

Jennifer, who was 9-years-old said:

'I like doing shapes and working with the circle and area and doing practical maths, things like getting a kick-wheel and measuring metres on a field or the playground and things like that.'

- Make the mystery of 'pi' work for you. Give them a surprise! Much has been written about pi and it has been computed to many decimal places. There are tables giving the value of pi to far more decimal places than any of our students could possibly remember... though it can be fun to challenge their memory. And they're often much better at it than we are.

This way they will remember that π is a number and that in practice when it is used it will only be approximate. As one teacher said:

'I'm not in favour myself of the loose use of pi as the number 3 and letting children do complete exercise taking pi as 3. I really do believe that from the beginning pi should be taken as 3 and a bit and it should be understood by children that it is approx. 3 and a seventh or 3.142.'

This teacher has a practical way of demonstrating that π is just a bit more than 3. He draws a circle and cuts four wires, equal, of a different colour and equal to the diameter ... three wires are then seen as not quite completing the circle.

- For older students, use π in formulae:

$C = \pi d$ or $\pi 2r$ or $2\pi r$, emphasizing that the order of multiplication does not matter. Stress the meaning of the symbols and how they are related, i.e. a 'length' on one side of an equation must be balanced by a 'length' on the other side (this might avoid the confusion between $2r$ and r^2).

- Use shapes to approximate the *area* of a circle. The Babylonians used squares (or hexagons) to touch inside and outside the circle and then halved the combined area of the inner and outer figures. The Greeks got nearer by using a regular polygon and gradually increasing the number of sides. To help establish πr^2, circles with radii in ratio 1:2 can be drawn on graph paper and the squares counted: areas will be seen to be in the ratio 1:4. Repeat for other circles. A neat way of establishing that $A = \pi r^2$ is to divide the circle into sectors and approximate each by a triangle. As the number of sectors is increased the approximation becomes more accurate and the sum of the bases approaches the circumference of the circle.

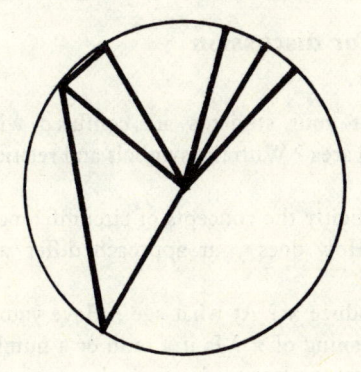

- For older students the work will become more formal and involve the use of the formulae $C = \pi d$ or $2\pi r$ and $A = \pi r^2$ or $\frac{\pi d^2}{4}$.

A secondary-school teacher:

'I think it's imperative that children should be able to manipulate equations, should be familiar with powers and be able to transpose symbols within a formula.'

And, repeating the words of the engineering professor (in Chapter 6), students should be able to

'...recognize an area formula ... just by looking at it in the sense of dimensions. After all, if there's one r or one d it must be a linear thing, the circumference of a circle for example. If there's an r squared or a d

squared, it must be the area, either the area of a circle or the surface area of a sphere.'

● And remember the role of the calculator in reinforcing a feeling and 'at-homeness' for the formulae; do not let awkward numbers obscure the concepts.

● Give your students a surprise, ask them this question:

'Imagine tying a rope round the world; it would have to be approximately 25 000 miles long. If 6 ft were added to this rope how far off the surface of the earth would it be?

Are they surprised? Is this why it's so difficult to tie a tight knot in the string when we pack a parcel?

Some questions for discussion

1. Do you find that your students are confused with the meanings of circumference and area? With the symbols and relations between them: π, r, d, C, A?
2. Do you teach explicitly the concepts of circumference and area? If so, at what age/ages? How does your approach differ with the age of your students?
3. How do you introduce π? At what age? Have your students acquired a feeling for the meaning of π? Is it a ratio or a number?
4. How much practical work with the circle do you encourage your students to carry out?
5. Is there a case for the use earlier on of area = 'circle number' × radius × radius to reinforce the concept of area as 'squaring' a dimension? When would you introduce the symbolism r^2 or d^2 to your students?
6. One aim of teaching maths should be 'to develop an understanding of mathematics through a process of enquiry and experiment' (*Mathematics 5–11*). Do you agree? How would you achieve this aim for students of all ages?

Guideline exercises

Try all these exercises with your students. They are a guide to constructing more of your own. The attractors are designed to reflect students' ways of thinking. Can you see what they are?

Guideline exercise 1

This exercise concentrates on formulae of circumference and area of a circle.

The circumference of a circle diameter d is?
a πd b πd^2 c $2\pi d$ d $\pi^2 d$ e

The area of a circle diameter d is?
a πd^2 b $2\pi d$ c $\pi\dfrac{d^2}{4}$ d πd e

The circumference of a circle radius r is?
a πr b $2\pi r$ c πr^2 d $2\pi + r$ e

The area of a circle radius r is?
a $2\pi r$ b πr^2 c $\pi^2 r$ d $(\pi r)^2$ e

Which of the following is true? The diameter of a circle is?
a twice the radius
b half the radius
c the same as the radius
d the radius squared
e

Guideline exercise 2

This exercise concentrates on finding the circumference and area of a circle.

With π as 3.14 the circumference of a circle diameter 10 mm is?
a 31.4 mm   b 62.8 mm   c 78.5 mm   d 314 mm   e

With π as 3.14 the area of a circle diameter 200 mm is?
a $(3.14)^2 \times 200$ mm^2
b 314 mm^2
c 628 mm^2
d 31 400 mm^2
e

The length of lace needed to go around the edge of a circular table mat of diameter 28 cm, with π as 3.14, will be?
a 616 cm   b 176 cm   c 88 cm   d 22 cm   e

The area of a circular disc of diameter 14 inches, with π as $\frac{22}{7}$, is?
a   88 sq inches
b   154 sq inches
c   176 sq inches
d   616 sq inches
e

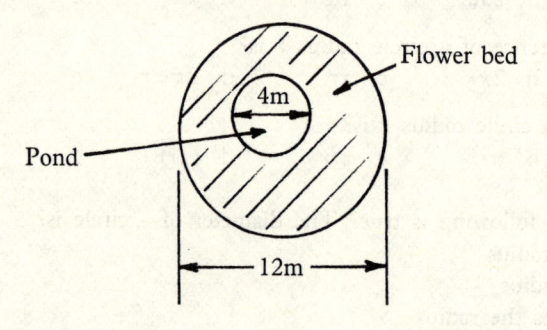

A circular flower bed of diameter 12 m, has a circular pond of diameter 4 m cut into it. The area of flower bed excluding the pond (i.e. the shaded area in the diagram) is?
a 8π m²   b 16π m²   c 32π m²   d 128π m²   e

With π as 3.14 the circumference of a circle radius 5 m is?
a 7.85 m   b 15.7 m   c 31.4 m   d 78.5 m   e

With π as 3.14 the area of a circle diameter 10 cm is?
a 314 cm²   b 78.5 cm²   c 62.8 cm²   d 31.4 cm²   e

The diameter of the base of a cone is 20 mm. The area of the base is?
a 10π mm²   b 20π mm²   c 100π mm²   d 400π mm²   e

The area of a circle radius 20 cm is?
a 400π cm²   b 100π cm²   c 20π cm²   d 10π cm²   e

Guideline exercise 3

This exercise concentrates on finding length, perimeter, area and volume given straight-line shapes.

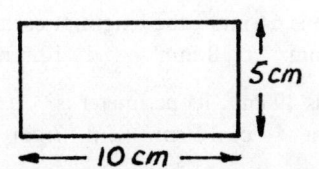

The area of a rectangle length 10 cm and width 5 cm is?
a 200 cm²   b 50 cm²   c 30 cm²   d 15 cm²   e

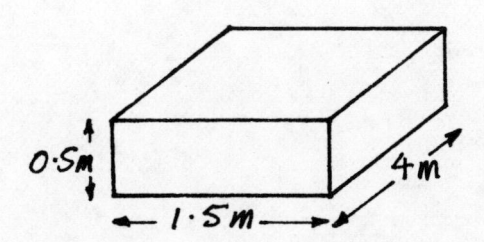

The volume of a solid length 4 m, depth 0.5 m and width 1.5 m is?
a 6 m³   b 4.75 m³   c 3.5 m³   d 3 m³   e

If I want to buy ornamental fencing to go around my 8 m by 5 m lawn, how long does it have to be?
a 80 m   b 40 m   c 26 m   d 13 m   e

A miniature painting 30 cm by 20 cm is mounted in the centre of a card so that there is a border of 5 cm of card around the outside of the painting. What is the area of the card?
a 1200 cm²   b 875 cm²   c 600 cm²   d 60 cm²   e

How many cubic metres of water must be pumped into a swimming pool 20 m long, 3 m wide, to raise the water level 2 m?
a 25   b 26   c 62   d 120   e

A rectangle has a length 12 m and a width 3 m. Another rectangle having the same area has a length of 9 m. What is its width?
a 3 m   b 4 m   c 6 m   d 7.5 m   e

If a square has an area of 36 cm², the length of each side is?
a 6 cm   b 9 cm   c 18 cm   d 36 cm   e

If the volume of a cube is 64 mm^3, the length of each side is?
a 4 mm   b 5.3 mm   c 8 mm   d 10.6 mm   e

If the area of a square is 49 cm^2, its perimeter is?
a 49 cm   b 28 cm   c 14 cm   d 7 cm   e

If the perimeter of a square is 20 m, its area is?
a 5 m^2 b 20 m^2 c 25 m^2 d 100 m^2 e

CHAPTER 8

AN INTRODUCTION TO ALGEBRA

'... the very essence of ...mathematical science... is the prevention of waste of the energies of muscle and memory.'

(Jourdain)

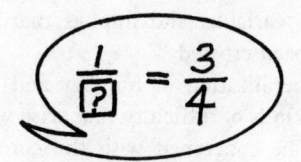

$\dfrac{1}{\boxed{?}} = \dfrac{3}{4}$

That's arithmetic ...
number in box is $\frac{4}{3}$

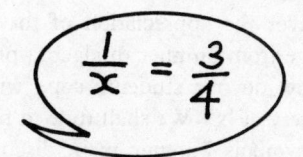

$\dfrac{1}{x} = \dfrac{3}{4}$

That's algebra... must be a formula ... can't remember

These were the impact perceptions of an undergraduate economics student and a headmaster ... and they were typical of many students' reactions to this task.

Is this how your students would react? Is this how you would teach it? In this chapter we are concerned with crossing the bridge between number and arithmetic and generalized number and algebra. Generalization and relations are the essence of mathematics and this surely is what algebra is all about. We have never subscribed to the view that number and arithmetic are a thing apart; our studies from the early 1970s have been based on the premise that developing a feeling for number is essential to the development of a feeling for symbolism and relations which are at the heart of algebra. If students clearly understand the notations and rules adopted for number, the ways in which digits are arranged to give place value, the rules when using the operators $\pm$, $\otimes$, the order of operations, etc., they will be better prepared to understand the generalization of these ideas with the use of symbols.

The *real understanding* of a relation such as $C = 5n$ where C stands for the cost of n articles at 5 units each can open students' eyes to the very essence of mathematics: the elegance of a language which is so concise and yet so general. And even further generalization with $C = kn$ where k is a constant ... how much effort this spares us (two pages of their own specific examples of $C = kn$ may persuade our students of the economy of mathematics!).

Building the bridge between number (arithmetic) and the generalized use of number (the beginnings of algebra) to the notion of the variable and finite processes (algebra) is a very skilled task and this is our job as teachers. Why should it be obvious that $2a$ means $2 \otimes a$ when 23 means 2 lots of ten $\oplus$ 3 lots of one?

We *must* be explicit when we introduce notation; the foundation for algebraic concepts must be solid. Understanding the symbols and the ways in which they can relate paves the way for the work we've been discussing with similarity and the circle. Understanding the meaning of and the use of formulae such as $C = \pi d$ or $2\pi r$; $A = \pi r^2$ or $\pi \frac{d^2}{4}$, $A \propto d^2$ and $V \propto d^3$ involves the appreciation of the notion of 'variable' and implies that the bridge from number to algebra proper has been crossed.

How do our students cope with the generalization of number and the mystery of 'x'? We shall tune in first to the kinds of difficulty that arise with the symbols but our main discussion will be concerned with the core of difficulty concept of the inverse '$\frac{1}{x}$'.

Diagnosis

Signs and symptoms

Let's now tune in to looking and listening.

Looking

Do you see this?

x ??
Help!
Not only
the
beginner!

$2 + 3x \ldots \boxed{5x}$

$2x + 3x \ldots \boxed{5x^2}$

$2x \cdot 3x \ldots \boxed{6x}$

Some common errors

x is 3,
x^2 is ...?
$\boxed{6}$
$2x^2$ is ...?
$\boxed{12} \ldots \boxed{36} \ldots$

An introduction to algebra 141

$$\frac{a}{2} + \frac{b}{5} \quad \ldots \boxed{\frac{a+b}{7}}$$

$$\sin x + \sin 2x \quad \boxed{\sin 3x}$$

$$\frac{1}{x} = \frac{3}{4} \quad x? \ldots \boxed{\frac{1}{12}}, \boxed{12}, \boxed{4}, \boxed{\frac{3}{4}}$$

Undergraduate engineers

All students: school, further education,
university, teacher trainees, TOPS

...and so on. You could no doubt quote many more examples. Let's look at some of the tasks we've set to students and the ways in which they see them.

10-year-olds were asked what number must go in the box to make the following true:

Task $12 + 6 - \square = 15$ | 87% correct

12-year-olds were asked to find x:

Task $12 + 6 - x = 15$ | 83% correct

The 10-year-olds found the next task rather more difficult.

Task $\square + 3 = 9 - 2$ | 64% correct
(20% chose 6
12% non-attempts)

10-year-olds

This next one introduced more variety!

Task Can you think of one number to put in the box and another to put in the circle so that it makes the following sum true?

$9 + \square = 5 + \bigcirc$ | 66% gave
11 different correct
answers
(13% non-attempts)

10-year-olds

12-year-old secondary-school students were set the following two tasks; compare the performances:

Task $17 - x + 7 = 20$ | 71% correct |

Task $8 + 7 = x + 5$ | 60% correct (21% chose 15) |

You will notice the drop in the success rate and yet we would surely think that the second task was easier than the first. Is this the effect of 'x on the other side'? Is students' thinking so rigidly attached to x on the left-hand side?

Let's move on to the work of older students in the secondary school.

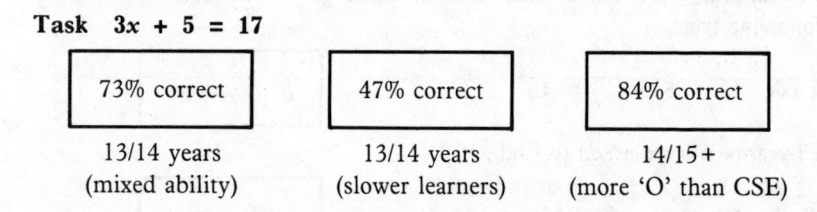

Task $3x + 5 = 17$

| 73% correct | 47% correct | 84% correct |
| 13/14 years (mixed ability) | 13/14 years (slower learners) | 14/15+ (more 'O' than CSE) |

Technician students on the first year of their course scored highly on the next task:

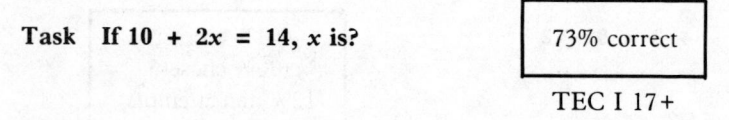

Task If $10 + 2x = 14$, x is? | 73% correct |

TEC I 17+

but look what happened with this one:

Task If $y + 6 = 2y - 7$, y is? | 54% correct (10% chose -13) (17% non-attempts) |

These students have a *minimum* entry requirement to their course of CSE grade 3. It is rather sobering to observe that one student in two could not find y and one in six did not even attempt it. It is probably one of those many maths tasks we can set where 'the bark is worse than the bite'. Nevertheless students such as these if they truly understood the concept of an equation

An introduction to algebra 143

Table 8.1
Task If $x = 5$, $y = 3$ then x^2y is?

Students	% choosing			% non-attempts
	Correct answer 75	28 (x^2+y)	30 ($2xy$)	
Secondary				
13/14 (mixed ability)	46	20	18	7
13/14 (slower learners)	15	30	21	6
14/15+ (more 'O' than CSE)	77	12	8	1
Craft/technician 17+	63		20	6
TOPS trainees (19–57 years)	33	20	19	18

Table 8.2
Task If $R = 4$, $I = 3$, then I^2R is?

Students	% choosing			% non-attempts
	Correct answer 36	13 (I^2+R)	24 ($2IR$)	
Secondary				
13/14 (mixed ability)	51	22		7
13/14 (slower learners)	22	16		7
14/15+	80	10		2
Craft/technician 17+	73		15	
Teacher-trainees	93			
Undergraduate engineers	99			
TOPS trainees	41	10		24

should have experienced no difficulty.

We found that the core of difficulty existed for these TEC (Technician Education Council) students just as it had for those City and Guilds of London Institute (CGLI) technician students tested during earlier years. Yet the evidence presented by the Technician Education Council to the Cockcroft Committee stated that technician students had no particular difficulty with mathematics. This assumption based on the normal TEC assessment procedures clearly demonstrates the *need* for diagnostic studies.

A different kind of task involving the understanding of symbolism was set to a variety of students. Look at their responses in Table 8.1. Notice the popularity of the responses which are the results of working out $(x^2 + y)$ and $2xy$. If you recall, it was this task and the students' confusion with the meaning of x^2 that led to the unravelling of one of the mysteries surrounding πd as a popular answer for the area of a circle. Now contrast the responses in Table 8.1 to those in Table 8.2 for the same task expressed with different symbols R and I, familiar to the craft and technician students and school physics students, as resistance and current in an electrical context.

It would be expected that students from about 14 years onwards would have met the symbols R and I as resistance and current respectively and may know that I^2R represents a heating effect. Were their higher success rates due to the task having more meaning for them? The effect of context will be discussed in Chapter 10 but before we jump to any general conclusion at this stage it must be emphasized that the *pure* context x^2y itself triggered confident responses, right or wrong, as will be seen later from the recordings.

- The inverse/'upside-down x'.

The tasks so far have been of the sort we traditionally set our students and they have been concerned with 'x on the top'. But what about 'x on the bottom' ... a position it occupies in so many contexts pure and applied? The difficulties that arise in the teaching of trigonometry can so often be traced to the unknown 'on the bottom'. And all those contexts involving an inverse variation, e.g.: speed, distance and time; density, mass and volume; resistance, potential difference and current; resistances in parallel; formulae in optics; pressure, volume and temperature. All these can involve an unknown on the bottom. And this is the despair of so many students, teachers and x!

An introduction to algebra

'I wish I were on top'

'I don't know what to do with x on the bottom'

Look at the performance of students in Table 8.3 on what might seem superficially a fairly straightforward task. The most popular incorrect response is $\frac{3}{4}$; a variety of other responses included 3, 4, 12, $\frac{1}{3}$ ... sometimes we wonder whether the choice depends on the day of the week, as in 'Monday we add, therefore Thursday we multiply!'

This is a core of difficulty task and there are several points to be noted from this table:

• The success rate of TEC students is virtually the same as that of the combined craft and technician students of the earlier studies but more of them

Table 8.3
Task If $\frac{1}{x} = \frac{3}{4}$, x is?

Students	% choosing		% non-attempts
	Correct answer $\frac{4}{3}$	$\frac{3}{4}$ ($\frac{1}{x}$)	
Secondary			
13/14 (mixed ability)	10	25	20
14/15 (more 'O' than CSE)	43	31	4
Craft/technician 17+			
Precourse	37	30	8
Postcourse	35	37	5
TEC (level I) 17+	36	23	18
TOPS trainees (19–57 years)	21	28	27

Table 8.4
Task If $\frac{1}{R} = \frac{1}{2} + \frac{1}{6}$ then R is?

Students	% choosing		
	Correct answer $\frac{3}{2}$ ($1\frac{1}{2}$)	$\frac{2}{3}$ ($\frac{1}{R}$)	8 (adding bottoms and inverting)
Secondary			
13/14	4	31	34
14/15+	25	36	29
Craft/technician 17+	16	52	11
Teacher trainees	36	52	5
Undergraduate engineers	66	33	0
TOPS trainees (19–57 years)	12	32	12

did not attempt the task.
● The jump in success rate between third-year and fourth-year secondary was typical of the results with similar tasks. What could you deduce from this?
● The success rate postcourse of the craft and technician students was virtually the same as that for precourse but more students attempted the task and chose wrongly. The pattern of performance pre- and post-course follows that for the other core of difficulty tasks, that is, performance on these tasks does not improve with the normal exposure to general mathematics teaching... *explicit* teaching is necessary.

Several tasks of this sort were set to a variety of students after we had studied the results from the very first one of this kind. These results are shown in Table 8.4 for comparison with those in Table 8.3.

In both Tables 8.3 and 8.4 the combined groups conceal differences. In the 14/15-years mixed-ability group, for instance, the 'O' level students scored significantly higher compared with the CSE students, as did technicians compared with the craft students. This is not to say that any scores were high. In fact on the $\frac{1}{R}$ task, the success rate for 'O' level students was only 37%.

The symbol 'R' may have been unfamiliar to the younger 13/14-year-olds; nevertheless, the jump in success rate between this group and 14/15-year-olds reflects the pattern with the $\frac{1}{x}$ task, suggesting more familiarity during the fourth year. This task was originally given to craft and technician students who were studying electricity as part of their course and were therefore familiar with R in this form, i.e. finding the effective resistance of two resistances in parallel. You will note how badly all students performed; even the undergraduate engineers could only boast a two thirds success rate. And the teacher trainees ... perhaps another point in the learning–teaching–learning cycle? Can you see, by comparing these two tables, what the blockage is? (Look carefully before you read on.)

You will have noted that the most popular incorrect response in Table 8.3 was $\frac{3}{4}$, i.e. $\frac{1}{x}$. From Table 8.4 you will see that the most popular incorrect response was $\frac{2}{3}$, that is the fractions have been correctly added but the answer has been left as $\frac{1}{R}$. We have drawn your attention to these tables in this way because the blockage appears to be the same for all the tasks we have set of this kind, that is, that even when students work out the fractions correctly they leave their answer in the form '$\frac{1}{x}$' because they do not know where to go from there. Yet take care before generalizing ... diagnosis is subtle.

Look at the next task:

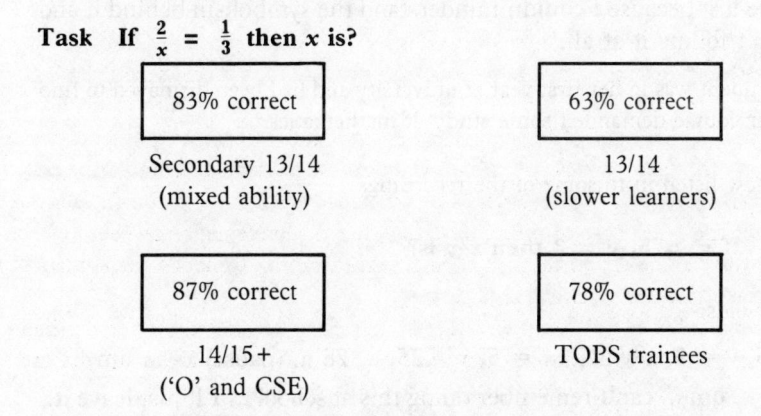

Task If $\frac{2}{x} = \frac{1}{3}$ then x is?

83% correct	63% correct
Secondary 13/14 (mixed ability)	13/14 (slower learners)
87% correct	78% correct
14/15+ ('O' and CSE)	TOPS trainees

and even for the younger 10-year-old students:

Task $\frac{2}{\square} = \frac{1}{3}$ **What goes in the box?**   54% correct and for the 12-year-olds:

If $\frac{2}{x} = \frac{1}{4}$ then x is? | 70% correct

Why should tasks which look so similar generate such wide differences in success rates? What do *you* think?

Let's now join our students in the language laboratory or eavesdrop on interviews.

Listening

Even the sight of an 'x' can put off both younger and older students from learning algebra.

Secondary-school student:

'I don't like algebra. I get muddled up with all the letters and numbers. I get them mixed up.'

Undergraduate marine biologist reflecting on her maths lessons at school:

'It was when all the xs and ys and equations came into maths that I became lost because I couldn't understand the symbolism behind it and couldn't follow it at all.'

This student was in her first year at university and had been dismayed to find that her course demanded some study of mathematics.

Let's now listen in to some of the recordings

Task   If $x = 5$, $y = 3$ then $x^2 y$ is?

13-year-olds:

'$x = 5$, $y = 3$, $x^2 y$ is ... $x = 5$, x^2...25, ... 28 ... (pause) x^2 ... um '.

'$x^2 y$ is ... um ... can't remember doing this at school ... I'll just leave it.' (Changes his mind and puts 13 into the space, i.e. $2 \times 5 + 3$)

17+ craft students:

'... x two y is ... 2 on top means x times x add y ... 28.'

'If $x = 5$, $y = 3$ then x^2y is, square of 5 is 25 plus $1y$ which is 3, 25 + 3 is 28.'

'x two y is ... 30 ... x equals 5, double amount that's 10, 10 threes are 30.'

'... 5 times 5 is 25 times y is 75.'

17+ technicians:

'... 5 squared times y which is 3 will be 25 times 3 ... 75.'

'25 times 3 equals 75.'

17+ ONC student:

'... 5 × 5 equals 25, y is 3 ... is 28. Oh! 3 × 25, 3 × 25 is 75.' (His first attempt was wrong but he recovered. Was it carelessness or the misconcept lingering from earlier days?)

Undergraduate engineers:

'x^2y is 25 plus 3 ... no 25 × 3 which is 75.'

'then x^2 plus y, x^2 equals 25, y equals 3, 3 × 25 is 75.'

'... x^2 is 5 × 5, 25 x 3 is 75.'

(Note that even some of these students stated *plus* at first, although they actually multiplied.)

On the whole working out x^2y is carried out confidently by most students even though their solutions may have been incorrect. It is the kind of task that 'triggers' a solution route whether right or wrong.

Listening in to the 'inverse' tasks, however, reveals a different reaction.

Task If $\frac{1}{R} = \frac{1}{2} + \frac{1}{6}$, R is?

13-year-old:

'1 over x equals a half plus 1 over 6. Now a half plus 1 over 6 you do ... . First of all you add the top numbers which is 1 plus 1 equals 2, then you add the bottom numbers which is 2 plus 6 equals 8... .'

Task   If $\frac{1}{x} = \frac{3}{4}$, x is?

17+ technician student:

S: Cross-multiply ... 3 times x = 4 times 1... x is $\frac{4}{3}$.
I: Why?
S: I don't know but it works.

(Cross-multiplication was a common technique with the technicians)

17+ craft students:

'If 1 = 3 ... typing looks funny. If 1 *over* ... oh $\frac{1}{x} = \frac{3}{4}$ then x ... 4 ... 1 is 3 so x must be 4.'

S: If one x = ...
I: Repeat the question
S: If one x = ...
I: Repeat the question
S: If one x = ...
I: What does that line mean?
S: 'Mm ... but I don't know what to do with x on the bottom.'

17+ ONC student:

'... then turning each one upside down ... both sides.'

This task does not trigger students into a favourite established routine, rather there is a searching around for a solution route. And the route they start on depends on their 'impact perception': some of the 13-year-olds, like the student above, 'saw' equivalent fractions but did not know how to proceed from there. The craft student could not get off the ground because he didn't know what to do with x on the bottom and the technicians cross-multiplied but didn't know why.

It is a task that generates anxiety and frustration and only the language-laboratory recordings can show this. Even university students do not solve such tasks completely automatically.

Undergraduate engineer:

'...dear, dear ... my mind is usual blank with this ... (pause) ... oh of course its ...um ... yes ... sorry ... I remember ... a moment to write it out ... new relationship ... $\frac{4}{3}$.'

We would like all our students to behave like this engineer ...

'... just a very simple... Multiply throughout by x and get $3x$ equals 4. Therefore x equals $\frac{4}{3}$.'

... perhaps they would if we taught *explicitly*.

Listen in to this interview with a 17+ student on a furniture course at a further-education college:

Task If $\frac{1}{x} = \frac{1}{4} + \frac{1}{3}$, x is?

S: 1 over x = a quarter plus 1 over 3.
I: What do you think the first step is to solve that question?
S: Add 1 over 4 to 1 over 3.
I: OK. And what do you think you'd get if you did that?
S: Well, I'd have to turn them into 12ths.
I: Yes, that's good.
S: So, I'd have 3 over 12 and 4 over 12.
I: Which is?
S: 7 over 12.
I: Excellent. So 7 over 12 equals 1 over x, can you solve that?
S: No.

This student was clear about the use of equivalent fractions to solve the addition but he came to grief at the inverse. The mathematics learnt at school does affect later studies, as this student said:

'As the course goes on, the maths gets much more advanced and I'm not sure whether I'll be able to complete the course if my maths isn't up to the standard required.'

And at school, maths has to be applied to other subjects and this blockage with the inverse can affect students' understanding in physics. For example, as this teacher said:

'Yes, it's a common problem and it causes difficulties nearly every time I meet it, and also not only with one ability of pupil. I've found difficulty with mixed-ability classes and with high-ability classes.'

This teacher also talked of her experience with the use of the formula $\frac{1}{f} = \frac{1}{v} + \frac{1}{u}$ for calculating focal lengths of lenses.

'Even if the distances are easy numerically, so that u and v are 2 and 3, we then have to add a half to a third and most people manage to find the answer five sixths, but often my problem begins there because we are then left with 5 over 6 equals 1 over f and the next stage of finding f seems to present more difficulty than adding the fractions.'

13-year-olds:

'$\frac{1}{x} = \frac{3}{4}$ then x ... three quarters, good. No. Must work this out, $\frac{1}{x}$ ... no can't work it out... its 1 something ... uh ... uh... do this one ... I think ... I think ... Oh, I don't know ... leave this one ... go back to this in a minute. No... In a minute.'

'... then $\frac{1}{x}$ is ... um ... um ... (long pause) ... huh (sighs) ... . If $\frac{3}{4} = 1$ ... then x is $\frac{1}{12}$ ... because ... because ... 3 ... 2 × 1 are 3 and 4 × 1 are 4, so if... oh ... $x = 1$.'

'... is ... (pause) ... x is um ... if you say 1 times by what is 3, which is 3, something times by 3 is 4, then that is $\frac{3}{4}$.'

(The last student was nearly there ... but we're into fractions!)

And what about the younger students?

Task Compare these two tasks: If $\frac{1}{x} = \frac{2}{4}$, x is? and
If $\frac{1}{x} = \frac{3}{4}$, x is?

13-year-old student:

'1 over x equals 2 over 4, so x equals 2 because 1 over 2 is a half and 2 over 4 is a half. x equals 2.

1 over x equals 3 over 4, x equals ... x equals ... (sigh) ... all right, x equals say 2.'

Actually *listening* to this recording as opposed to reading it is an eye-opener.

The next solutions to $\frac{1}{x} = \frac{3}{4}$ will encourage us (in one task x was replaced by a box):

'Well if you turn 1 over x upside down, you get x over 1 which equals 4 over 3 and 4 over 3 is 1 and a third. Therefore x equals 1 and a third.'

'1 over, and then it's got the box, equals three quarters. And I work it out because you multiply the 1 by 3 and so to get the answer in the box you divide 4 by 3, which is, 3s into 4 ... is 1.3 recurring.'

(Note the different ways in which they solved the task.)

Here are some interviews with 10-year-olds about filling in the box.

Task $\frac{2}{\Box} = \frac{1}{3}$

First student:

S: 2 over something equals one third.
I: Yes, now I want you to find what the something would be. What would I put in that box?
S: 6.
I: Why do you say that?
S: Because half of 6 is 3 and in fractions the bigger number is lower than the smaller number.

Second student:

S: (Reads the task aloud)
I: Have you seen this type of sum before?
S: Yes ... 6.
I: Why?
S: Cos that ... 2 is double 1 ... so 6 is double 3.

But when he came to do this one:

Task $\frac{\Box}{8} = \frac{3}{12}$

S: Well ... 8 goes into ... that ... well ... (laughs) ... sorry ... you times that by ... you've added 4 to that ... ohh! ... hold it I'm stuck! ... ahh! ... (laughs) ... right! so ... that! ... you've added 4, cause 8 goes ... 4 and 8 ... I mean 4 goes into 8 and 12 so ... so ... so ... it goes ... it goes ... mmmmm ... so 4 goes three times into that ... oooooh.
I: You're doing well. Keep trying.
S: So it goes 2 into 8 so I think it will be 2.

All these tasks illustrate the subtlety of diagnosis: they look the same yet the impact perceptions and the routes to solution which they generate are quite different.

How do teachers perceive $\frac{1}{x} = \frac{3}{4}$?

'I see 1 over x = 3 over 4 as an equation and I would cross-multiply to get $3x$ equal to 4, and then divide the 4 by the 3 to obtain the value of x.'

'and I'd see the problem as an equation to be solved for x. To do that I'd reciprocate both sides to get x = 4 over 3.'

'1 over x is 3 over 4. All you have to do is turn them upside down, so you get x is 4 thirds. That's it.'

'My instant reaction is therefore x equals 4 over 3, I see it as reciprocal and I have to turn the 3 over 4 upside down to get 4 over 3.'

and this was a reaction of a professor of education (Professor Skemp in *Maths with Meaning*):

'Initially I see it as an equation of two reciprocals, but if I went on, I think I would see it as an equivalence of two ratios. I might see it as a fractional equation, and if I thought a bit longer I might see it in other ways as well. There's three to be going on with anyhow.'

What can we learn from all this looking and listening?

Learning

First, what can we learn about the early stages of introducing our students to algebra? We saw that a certain rigidity can quickly set in if all our exercises are set with the unknown on the left-hand side. Shifting the 'boxes' or the xs around is itself a generalization as is varying the symbols. So let's be flexible and encourage flexibility in our students.

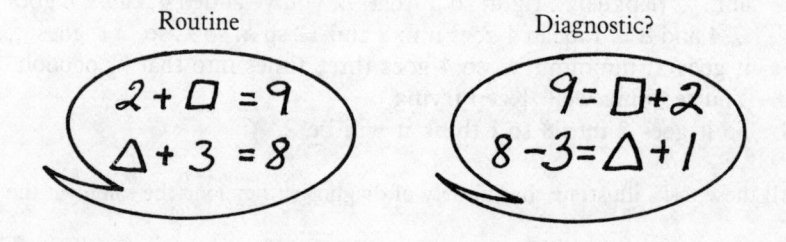

Routine

$2 + \square = 9$
$\triangle + 3 = 8$

Diagnostic?

$9 = \square + 2$
$8 - 3 = \triangle + 1$

An introduction to algebra 155

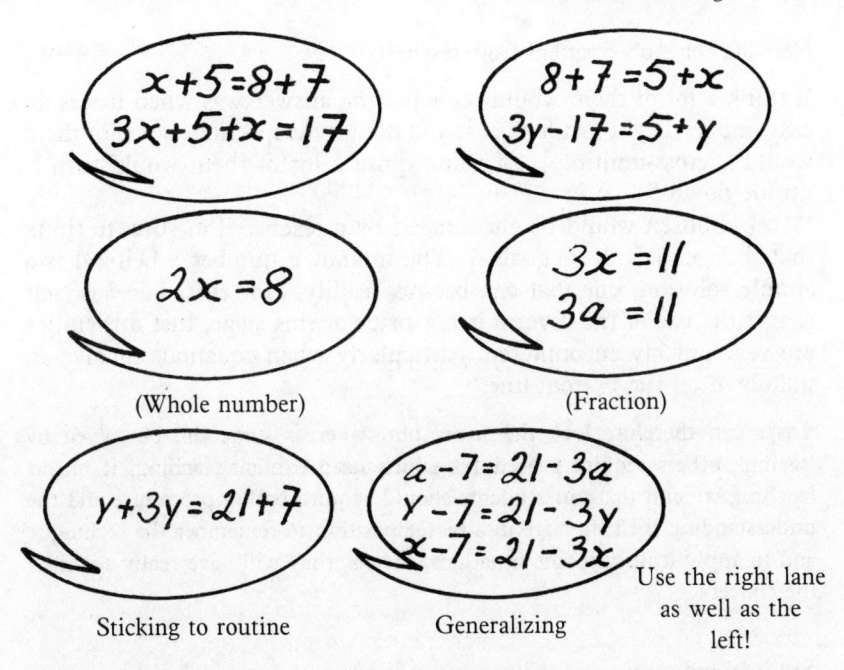

$$x+5=8+7$$
$$3x+5+x=17$$

$$8+7=5+x$$
$$3y-17=5+y$$

$$2x=8$$

$$3x=11$$
$$3a=11$$

(Whole number)   (Fraction)

$$y+3y=21+7$$

$$a-7=21-3a$$
$$y-7=21-3y$$
$$x-7=21-3x$$

Sticking to routine   Generalizing   Use the right lane as well as the left!

These are only 'tasters', of course, to which from your own experience you will be able to add many more. But changing the variable and its position will move our students on from the generalized arithmetic 'think of a number' procedure to the more general processes of algebraic thinking. And this is perhaps what we ought to do with the inverse.

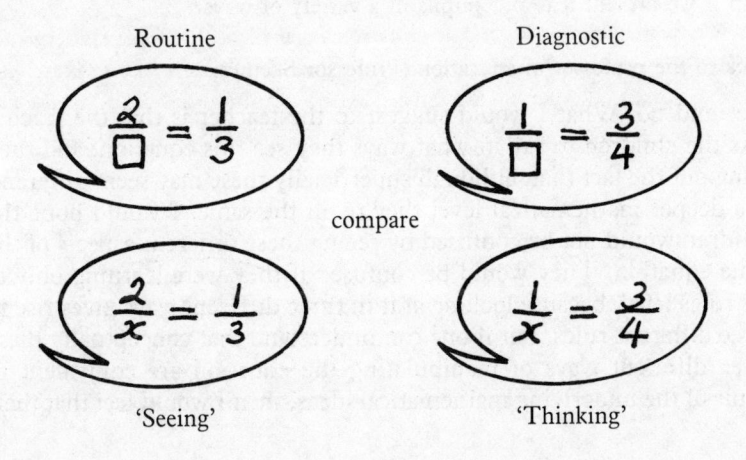

Routine   Diagnostic

$$\frac{2}{\Box}=\frac{1}{3}$$

$$\frac{1}{\Box}=\frac{3}{4}$$

compare

$$\frac{2}{x}=\frac{1}{3}$$

$$\frac{1}{x}=\frac{3}{4}$$

'Seeing'   'Thinking'

Here are some observations from teachers:

'I think a lot of them would see what the answer was when it was an easy one ... they'd just look at it and do it by inspection. Some of them would... cross-multiply ... I think quite a lot of them would turn it upside down.'

'Most children would be encouraged by a teacher, I'm sure, to think that if $2x$ equals 8, x equals 4. The unknown number x being 4 is a simple solution, one that can be *seen* readily. However, I feel myself that if the use of the inverse is not made at this stage, that difficulties are very quickly encountered, particularly when equations involve an unknown on the bottom line.'

Tasks can therefore look the same, but, whereas some can be solved by 'seeing', others require procedures which need explicit teaching. If in our teaching we aim that our students should acquire both a procedure and the understanding then they are in a better position to remember the technique and to apply it in different situations, that is, they will have really acquired the concept.

Student and teacher: is there a mis-match?

We have glimpsed the different ways in which students initially see a task of the form $\frac{1}{x} = \frac{3}{4}$ and the different ways in which the teacher may see it. This could clearly result in confusion. We must, however, sieze the opportunity that this variety provides to teach for flexibility and generalization. Does it help if we present it to our pupils in a variety of ways?

Back to the professor of education (Professor Skemp):

'Yes and no. What I would suggest to the teacher is that the teacher asks the children to say in what ways they see this equation, and then bring out the fact that, although superficially these may seem different, at a deeper mathematical level they're all the same. I would hope the children would not be confused by seeing these different aspects of the same equation. They would be confused if they were learning only at the rules level, because looking at it in three different ways gives rise to three different rules, but if one can understand that conceptually these three different ways of manipulating the equation are equivalent in terms of the underlying mathematical ideas, then I would feel that their understanding had been enlarged, and so far from being more confused they would have gained an understanding.'

and the teacher of physics:

'I usually find I have to attack it from several angles before I've convinced a class of people. So I'll look for a flow-diagram method. I'll look for multiplying both sides by the same thing, I'll look for the reciprocal method which immediately occurs to me.'

Different ways of tackling a task should be backed up with plenty of practice so that the solution routes apparently become routines... but routines with *meaning*.

Professor Skemp again:

'I would see routinizing as having an important function at all levels of mathematics, from the solution of equations, the manipulation of formulae, right back even, ...yes, even to learning multiplication tables. But the routine processes must never lose their connections with the underlying mathematical knowledge. Otherwise, they become rules without reasons and people are using the rules without understanding and this means that they may well use the rules inappropriately.'

Can we then pinpoint the diagnosis?

The diagnosis

Misconcepts associated with a lack of feeling for inverse relations involving $\frac{1}{x}$.

Prescription and aftercare

Suggestions

Introduce your students to the feeling for the inverse. The inverse is a mathematical way of getting back to where we were: $\ominus 3$ is the inverse of $\oplus 3$ and vice versa, $\oslash 3$ is the inverse of $\otimes 3$ and vice versa. It is the latter kind in which we are most interested and this is usually called the reciprocal, i.e. $\frac{1}{2}$ is the reciprocal of 2 (2 is the reciprocal of $\frac{1}{2}$), $\frac{1}{3}$ is the reciprocal of 3 (3 is the reciprocal of $\frac{1}{3}$) and generalizing, $\frac{1}{x}$ is the reciprocal of x.

158 Diagnosis and prescription in the classroom

In everyday life, if two quantities vary inversely, then the larger one gets, the smaller the other becomes and vice versa: e.g., greater speed, less time on a given journey, more pressure less volume at a given temperature, etc.

As a professor of engineering said:

'Well, the engineer is always concerned with magnitude and relative magnitude. When I look at 1 over x my first response is that if 1 over x is less than 1, obviously x is greater than 1, or to generalize if x is very large, 1 over x is very small and vice versa. So my approach is that of an inverse relation.'

And you can use this approach as you did with large and small numbers in the early chapters.
- From x very large to x very small.
- Use the overflow on the calculator.
- Range from the infinitesimal (inside atoms) to the infinite (on the planets).

Generalization

Formalize the material in preparation for solution of tasks in ways appropriate to the needs of your students:
- For younger students generalize with use of $\Box$, $\triangle$, $\bigcirc$, ... for the unknown and move it about in its position in a relation.
- For older students generalize with the use of different symbols $a, b, c, ..., x, y, z$ and move these about in relations.
- Make explicit the meaning and use of symbols.
- Ensure that students can work with equivalent fractions and add fractions ... the fraction concept is always popping up!

Flexibility in solutions

Ways of solving tasks such as $\frac{1}{x} = \frac{3}{4}$
- Reciprocal

$\frac{1}{x} = \frac{3}{4}$
$\frac{x}{1} = \frac{4}{3}$

The reciprocal leads to the rule 'turn both sides upside down' but remember to add any fractions first as in $\frac{1}{x} = \frac{1}{2} + \frac{1}{4}$ (say)

- Equivalent fractions

  $\frac{1}{x} = \frac{3}{4}$

  $x = \frac{4}{3}$

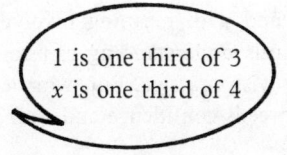

  (1 is one third of 3; x is one third of 4)

- Proportion (two equal ratios)

  $\frac{1}{x} = \frac{3}{4}$ (if $\frac{a}{b} = \frac{c}{d}$ then $bc = ad$)

  $3x = 4 \otimes 1$

  $x = \frac{4}{3}$

  ($\oplus$ both sides by 3)

- Equation

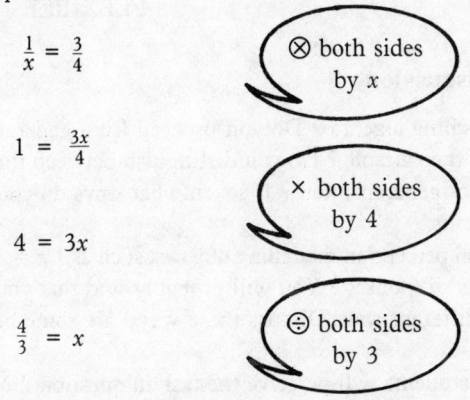

  $\frac{1}{x} = \frac{3}{4}$ *($\otimes$ both sides by x)*

  $1 = \frac{3x}{4}$ *($\times$ both sides by 4)*

  $4 = 3x$

  $\frac{4}{3} = x$ *($\oplus$ both sides by 3)*

And of course we could stop here and talk about equations but we haven't the time or space. So just one heartfelt suggestion ... treat an equation as a balance, think of it physically, use Dienes' balance if you have one.

Hence, whatever operation you apply to one side, be fair and do the same to the other: be explicit about this and please *don't* 'change sides, change signs'!

Practice

Last but by no means least ... lots of practice! To work through a solution correctly, confidently and with meaning involves both understanding and practice. For some pupils understanding can come first but for others the 'penny will drop' only with practice. But, whichever comes first, familiarity coupled with success breeds confidence and skill.

Learners (& *x*)   'Sitting on top of the world'   Teachers (& *x*)

CONCEPTS COMPETENCE
CONFIDENCE

AWARE EXPLICIT
FLEXIBLE

Some questions for discussion

1. What is your aim in teaching algebra? Do you proceed from generalized number to the notion of the variable? Do you distinguish between these? Is algebra the language of generalization? If so, in what ways do you get this across to your students?
2. What was your own initial perception/modelling of a task such as $1/x = 3/4$? What are your colleagues' responses? You will probably find that you are each perceiving it in a different way. Discuss these ways: are some better than others?
3. How do you think your students will perceive the task in question 2? Ask them. See if it matches up to your prediction. Does it matter if there is a mis-match between teacher and learner in the way each perceives the task? What implication does a mis-match have for your teaching?
4. Psychologists say that abstraction of a concept requires several experiences of that concept. Do you expose your students to several ways of solving a task? If not, why not? Do you think they would become confused? Are you clear in your own mind about different kinds of solutions?
5. A physics teacher when asked what she thought was the most important thing for her students to be able to do mathematically in their physics lessons said, 'Obtain a quick answer.' Discuss this in the light of the following quotation: 'One general aim of teaching maths should be to

An introduction to algebra 161

develop mathematical skills and knowledge accompanied by the quick recall of basic facts' (*Mathematics* 5–11). Do you agree? How would you set out to achieve this aim?

Guideline exercises

Try all these exercises with your students. Remember they are guideline exercises from which you can construct more of your own. The attractors are designed to reflect students' ways of thinking.

Can you see what they are?

Guideline exercise 1

This exercise concentrates on substitution.

If $p = 3$ and $q = 4$ then p^2q is?
a 144   b 36   c 24   d 13   e

If $s = 2$ and $t = 3$ then st^2 is?
a 11   b 12   c 18   d 36   e

If $x = 4$, $y = 0$ and $z = 5$ then xyz is?
a 0   b 5   c 9   d 20   e

If $v = \sqrt{3gh}$ one possible value of v, when $g = 8$ and $h = 6$, is?
a 12   b 72   c 144   d $\sqrt{386}$ e

If $x = 4$, $y = 0$ and $z = 5$, then the value of $xy + z$ is?
a 45   b 20   c 9   d 5   e

Do you recognize these contexts?

If $I = 4$ and $R = 3$, then I^2R is?
a 19   b 24   c 48   d 144   e

If π is taken as 3 and r is 5, then πr^2 is?
a 225   b 75   c 30   d 28   e

If $t = 2\pi\sqrt{\dfrac{l}{g}}$, then a value for t when π is taken as 3, $g = 10$ and $l = 90$ is?
a 9   b 18   c 27   d 54   e

If $y = mx + c$, the value of y when $m = 6$, $x = 0$ and $c = 5$ is?
a 0   b 5   c 11   d 65   e

If $v = u + at$, the value of t when $u = 4$, $a = 2$ and $v = 10$ is?
a 16   b 7   c 4   d 3   e

Guideline exercise 2

This exercise is concerned with the solution of equations.

What must go in the box to make the following true?
$\square + 4 = 11 - 2$
a 5   b 7   c 9   d 13   e

What must go in the triangle to make the following true?
$\triangle + 5 = 11 - 6$
a 12   b 6   c 5   d 0   e

What must go in the circle to make the following true?
$5 + \bigcirc = 12 - 8$
a 15   b 7   c 4   d −1   e

What number must z be to make the following true?
$9 + 8 = z + 6$
a 5   b 11   c 17   d 23   e

What number must x be to make the following true?
$10 - 8 = x - 1$
a 19   b 3   c 2   d 1   e

If $5x = 20$ then x is?
a 25   b 20   c 15   d 4   e

If $3x = 13$ then x is?
a 16   b 13   c 10   d $4\frac{1}{3}$ e

If $10 + 2p = 28$ then p is?
a 8   b 9   c 16   d 18   e

If $3p + 1 = 14$ then p is?
a $4\frac{1}{3}$ b 5   c 13   d 15   e

If $y + 7 = 2y - 8$ then y is?
a −15   b −1   c 5   d 15   e

Guideline exercise 3

This exercise concentrates on the $\frac{1}{x}$ task.

$\frac{2}{\Box} = \frac{1}{5}$
a $\frac{2}{5}$ b $\frac{5}{2}$ c 5   d 10   e

If $\frac{2}{x} = \frac{6}{24}$ then x is?
a $\frac{1}{4}$ b 8   c 12   d 24   e

If $\frac{2}{x} = \frac{5}{20}$ then x is?
a 40   b 20   c 8   d 4   e

If $\frac{1}{x} = \frac{5}{8}$ then x is?
a $\frac{8}{5}$ b 8   c 5   d $\frac{5}{8}$ e

If $\frac{3}{x} = \frac{2}{5}$ then x is?
a 6   b $\frac{15}{2}$ c 5   d $\frac{2}{15}$ e

If $\frac{1}{R} = \frac{1}{2} + \frac{1}{4}$ then R is
a $\frac{3}{4}$ b $\frac{1}{6}$ c 6   d $\frac{4}{3}$ e

If $u = 6$, $v = 2$ and $\frac{1}{f} = \frac{1}{u} + \frac{1}{v}$, then f is?
a 8   b $\frac{8}{2}$ c $\frac{3}{2}$ d $\frac{2}{3}$ e

If $\frac{3}{R} = \frac{1}{6} + \frac{1}{3}$ then R is?
a 27   b 6   c $\frac{1}{2}$ d $\frac{1}{6}$ e

A formula may be written as speed $= \frac{\text{distance}}{\text{time}}$ or $S = \frac{d}{t}$

If $S = 5$ and $d = 200$ then t is?
a 40   b 195   c 205   d 1000   e

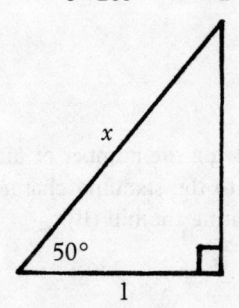

From the figure $\cos 50° = \frac{1}{x}$ then x is?
a $\cos 50°$ b 1   c $\sin 50°$ d $\frac{1}{\cos 50°}$ e

Guideline exercise 4

This exercise concentrates on the forming of equations.

A company calculates the cost of an article (A) by adding the cost of production (C) to the intended profit (P). Which one of the following would be the correct formula for calculating the cost of production C?

a   C = P + A
b   C = P − A
c   C = A + P
d   C = A − P
e

Take-home pay (P) is worked out by deducting tax and national insurance (D) from gross pay (G). Which one of the following would be the correct formula for working out take-home pay (P)?

a   P = G − D
b   P = G + D
c   P = D − G
d   P = D + G
e

The percentage increase in sales (P) is worked out by subtracting the sales for last year (L) from the sales for this year (T) and calculating this figure as a percentage of L. Which one of the following would be the correct formula for working out the percentage increase (P)?

a $P = \frac{L}{T} \times 100\%$
b $P = \frac{T-L}{L} \times 100\%$
c $P = \frac{T}{L} \times 100\%$
d $P = \frac{L-T}{L} \times 100\%$
e

A bill is calculated by multiplying the number of units used (U) by the cost per unit (C) and adding this to the standing charge (S). Which one of the following is correct for calculating the bill (B)?

a   B = U + C + S
b   B = UC + S
c   B = UCS
d   B = U + CS
e

A gas bill is calculated by multiplying the number of units used (U) by the cost per unit (in pence, C) and adding this to the standing charge (in pounds, S). Which one of the following would give the correct gas bill (B)?

a $B = U + C + S$
b $B = UC + S$
c $B = 100UCS$
d $B = \dfrac{UC}{100} + S$
e

CHAPTER 9

STATISTICS: ON AVERAGE

'There are three kinds of lies: lies, damned lies and statistics.'

(Disraeli)

'Statistical thinking will one day be as necessary for efficient citizenship as the ability to read and write.'

(H.G. Wells)

The term 'average' occurs frequently in everyday life: average wage, average height, average weight, the average family, Mr Average and so on. What does the word average tell us in each of these examples? What are we to understand from the term average? What does it mean when we are told 'the average working man earns £120 a week'? How was this 'fact' determined?

We loosely use the word average to describe a representative value. There is more than one way of selecting a representative. Clearly the method is important. Consider the following situation: a company has 15 employees; ten earn £85 a week while the remaining five earn £175 a week. What is the average pay per week? £85 (i.e. the modal/median wage), £130 [i.e. (85 + 175)/2] or £115 (i.e. the mean wage)? Our answer will depend on the arguments that we want to support.

The variety of names (e.g. mean, median and mode) under the umbrella term average can provoke confusion. Where do the misconcepts arise and what can we do about them?

Did you know that half the population is below average height?

Diagnosis

Signs and symptoms

Looking

- The basic idea: the arithmetic mean (for an example see Table 9.1).

Table 9.1

Task The average of 15, 17, 18.25, 15.75 is?

Students	% choosing			% non-attempts
	Correct answer 16.5	Total 66	Highest 18.25	
Secondary				
13/14	49 **24**	11 **15**	10 **16**	5 **13**
14/15+	77	6	10	3
TOPS trainees (19–57 years)	56	4	4	11

Numbers in **bold type** refer to slower learners.

Table 9.2

Task The weekly wages of four teenagers are £15.00, £17.00, £18.25 and £15.75. The average weekly wage is?

Students	% choosing			% non-attempts
	Correct answer £16.50	Total £66	Last wage £15.75	
Secondary				
13/14	54 **31**	12 **21**	9 **19**	4 **5**
14/15+	80	5	6	3
TOPS trainees (19–57 years)	70	12	3	5

Numbers in **bold type** refer to slower learners.

168 Diagnosis and prescription in the classroom

Does context affect performance? Look at the next task shown in Table 9.2. Context here has helped to improve performance. Although we used the word *average*, what is the reaction to the use of the word *mean?* In a recent national examination of 17+ candidates a question was posed asking for the mean time taken to get to work over a few days. The correct answer was given by 52% of the candidates, 24% gave the mode and 18% gave the median.

The introduction of a frequency table to simplify the data can provoke even more confusion. Consider Table 9.3.

Table 9.3 Frequency of broken biscuits in packets of chocolate fingers

No. of broken biscuits	Frequency
2	5
4	6
6	7
9	2
11	1

Task What is the mean of the distribution?

TEC level II students gave the following responses:

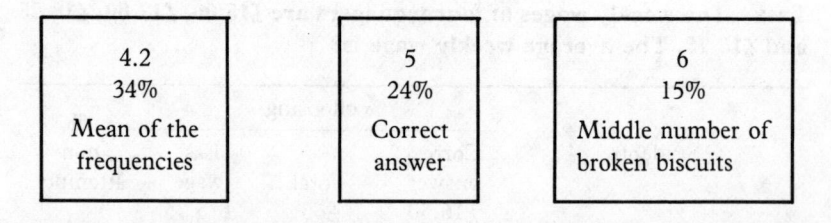

4.2	5	6
34%	24%	15%
Mean of the frequencies	Correct answer	Middle number of broken biscuits

Each of these answers is feasible as a representative of the typical number of broken biscuits, but the answers of 4.2 and 6 suggest a lack of understanding of:

The rationale behind a frequency table
How the mean is determined given a frequency table
The purpose of finding an average.

- Different types of average.

Median

of 5, 7, ⑥, 3, 4 ?

Mode

of 9, 8, 7, 9, 7, 9, 7, 8, 9 ?

Median... the middle value

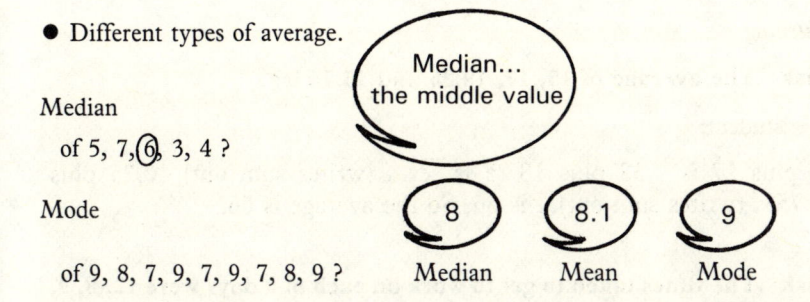

| 8 | 8.1 | 9 |
| Median | Mean | Mode |

In a recent national examination 17+ candidates were asked to state the modal time taken to get to work given the times for a number of days. The correct answer was given by 55% of the candidates, 22% gave the median time and 18% gave the mean time.

If a frequency table is introduced the concepts of median and mode can be probed quite deeply. Consider Table 9.3.

Task   What is the median of the distribution? (Refer to Table 9.3)

4	5	6	7
24%	15%	30%	15%
Correct answer	Median of frequencies	Median of the no. of broken biscuits column	Middle value in the frequency column

The idea that the median is the middle value has clearly been grasped here. Middle value of *what* is not so clear.

Task   What is the mode of the distribution? (Refer to Table 9.3)

6	7	11
56%	21%	7%
Correct answer	Highest value in the frequency column	Highest value in the no. of broken biscuits column

The idea that the mode is the most popular/most common/highest value appears to be understood. Most popular value of *what* is not so clear.

Listening

Task The average of 15, 17, 18.25 and 15.75 is?

13+ student:

15 plus 17 is... 32 plus 18.25 is...er... (writes sum out) 40.25 plus 15.75... (writes sum out)... is 66. So the average is 66.

Task The times taken to get to work on each of 9 days were 12, 8, 9, 12, 10, 12, 11, 8 and 12 minutes. The median time taken is?

17+ student:

I: How do you find the median?
S: Find the middle number.
I: Middle number... so what's your answer?
S: Er... ten. Ten is the middle number.
I: So what is the median?
S: Ten.

Task (Referring to Table 9.3) What is the mode of the distribution?

TEC level II student:

I: How do you find the mode?
S: Its something to do with the biggest number.
I: What is the mode of this distribution?
S: Seven.
I: How did you get seven?
S: Its the biggest number in that column (points to the frequency column).

Learning

'Surface concepts' appear to have been learned. Oversimplified rules are used, for example, to find the

Mean: add up all the numbers and divide by the number of numbers
Median: look for the middle value
Mode: look for the highest number.

The rules if used with understanding will give the correct answers. Applying the rules depends on the depth of understanding of the basic idea of average. Confusion clearly exists as to how to apply the rules outlined above. Why should such confusion exist? What can be done to elaborate on these rules so that they are applied with understanding?

The diagnosis

The general diagnosis is deficiency in a feeling for what an average is. It is important to:

● Know the reasons for finding an average.

For example, when we talk about average wage are we referring to mean wage? modal wage? Why do we use one average rather than another?

● Understand the result of finding an average.

For example, the average family has 2.45 children (?). Have you seen 0.45 of a child? What does this mean (!)? How will we use this information?

● Know a clearly defined method of finding an average.

For example, median is the middle value... of what? We must be clear.

Prescription and aftercare

Aim To enhance our students' development
of CONCEPTS COMPETENCE and CONFIDENCE.
by ourselves being AWARE EXPLICIT and FLEXIBLE.

It is essential that we make it clear from the beginning that we use the term average to refer to a representative from a collection of data. Discussion of how to select a representative can stimulate students to realize the difficulties in deriving a method that is applicable to a variety of different kinds of data.

The use of contexts in discussions will help reinforce the need for different kinds of average. Classical examples of average wages, stock control (in the shoe shop), test scores, rainfall, speeds, cricket scores and so on are all useful topics for discussion.

When we pose tasks in context we should be explicit about the required goal. It is not a good idea to ask for the mean, given a number of heights, say. We need to ask for the *mean height*. We should attempt to direct our students to a meaningful answer.

Many exercises simply ask for a calculation. A useful further question is: what *use* could be made of the result? For example, what can be made of the average wage, average mark, average rainfall and so on? This idea could be extended if a group of students could be split in two; they could be given information about wages and asked to present a case for wage negotiations from a management versus non-management point of view. Another exercise for groups of teachers/students is to look at a non-contextual example and invent a 'real' situation to match the data and put the result to practical use.

Once we have practised our students in the methods of finding an average it is a good exercise to ask them to read a description of a method used by a factory worker/piece worker/weatherman/stocktaker and so on to find an average and ask them to distinguish which method has been used. For example, a factory worker was asked to count the number of components she made each hour for a total of 30 hours. She then had to work out her average rate by dividing by 30 hours. This type of average is?

Some questions for discussion

1. The subject of statistics lends itself to being taught in context. List the relevant contexts that you feel would be 'realistic' to your students for finding an average. What are familiar contexts for your students? Test the contexts you would like to use. Is there a mis-match?

2. When should you introduce the idea of mean, median and mode? Should you restrict your early teaching averages to the mean only? If so, when should you introduce the median and the mode?

3. 'The idea of a frequency table as a means of summarizing and simplifying a collection of data appears to obscure the "sense" of the data for many students'. Do you agree? If so, is this due to a lack of practice in constructing frequency tables? Should we set the exercise of collecting and tabulating data more often, to help develop a better understanding of a frequency table?

4. How do you think your students perceive the tasks of finding the *a* average *b* mean *c* median *d* mode of a set of data? Ask your students to explain how to find these averages.

5. Is there a mis-match between the teacher and the student in the way each perceives the tasks of question 4? What implications does this have for teaching?

Guideline exercises

Try all these exercises with your students. They are a guide to constructing more of your own. The attractors are designed to reflect students' ways of thinking. Can you see what they are?

Guideline exercise 1

This exercise concentrates on finding the mean, median or mode of data not in a frequency table.

What is the average of 7, 5, 9, 11?

a 7   b 8   c 11   d 32   e

What is the mean of 8, 8, 8, 9, 9, 10, 10, 13, 15?

a 8   b 9   c 10   d 90   e

What is the median of 10, 10, 8, 9, 13, 15, 9, 8, 8?

a 13   b 10   c 9   d 8   e

What is the mode of 9, 10, 8, 9, 13, 8, 15, 8, 10?

a 8   b 9   c 10   d 15   e

What is the median of 2, 100, 1, 7, 100?

a 100   b 42   c 7   d 1   e

Each of the above tasks could be put into a context. For example:

The marks obtained in four spelling tests, each out of 12 marks, were as follows: 7, 5, 9, 11. What is the average mark?

a 32   b 11   c 8   d 7   e

The amount of rainfall over 9 consecutive days in Manchester was recorded as 8, 8, 8, 9, 9, 10, 10, 13, 15 mm. What was the mean rainfall in mm during these 9 days?

a 90   b 10   c 9   d 8   e

The time taken to get to work on each of 9 days was 10, 10, 8, 9, 13, 15, 9, 8, 8 minutes. The median time in minutes taken was?

a 13   b 10   c 9   d 8   e

The annual salary of seven employees in a company are £5,000, £37,000, £17,000, £15,000, £7,000, £12,000, £5,000. The median wage is?

a £15,000   b £14,000   c £12,000   d £5,000   e

During an hour a shoe-shop salesman sells shoes of the following sizes: 7, 8, 7, 6, 9, 8, 10, 9, 8, 8. What is the modal shoe size?

a 10   b 9   c 8.6   d 8   e

Guideline exercise 2

This exercise concentrates on finding the mean, median or mode for data presented in a frequency table.

The information below relates to the number of shoes sold during a lunch hour.

Shoe size	Frequency
4	4
5	2
7	2
8	3
9	5
10	1

The median shoe size sold during the lunch hour is?

a 9   b 8   c $7\frac{1}{2}$ d $2\frac{1}{2}$ e

The modal shoe size sold during the lunch hour is?

a 4   b 5   c 9   d 10   e

The mean shoe size sold during the lunch hour is?

a 2.83   b 7   c 7.16   d 9   e

Discuss these results.

CHAPTER 10

PROBLEM SOLVING: CONTEXT, ATTITUDE AND ANXIETY

'Facts are sometimes quite useful'.

(Jourdain)

'It is foolish to answer a question that you do not understand. It is sad to work for an end that you do not desire'.

(Polya)

Polya's statement, taken from his book *How to Solve it*, would seem to apply to so many students when faced with mathematics tasks in context. Understanding and motivation in our students are desirable objectives for us to keep in mind when seeking relevant contexts for mathematics. However, understanding the context and motivation to solve the problem even together are not sufficient. Jourdain's succinct statement, from the *Nature of Mathematics*, reminds us that mathematical knowledge is also a requirement for successful problem solving.

But what is problem solving? Is it important? The Cockcroft Committee Report *Mathematics Counts* says:

'The ability to solve problems is at the heart of mathematics. Mathematics is only 'useful' to the extent to which it can be applied to a particular situation and it is the ability to apply mathematics to a variety of situations to which we give the name 'problem solving'.'

(Paragraph 249)

'All children need experience of applying the mathematics they are learning both to familiar everyday situations and also to the solution of problems which are not exact repetition of exercises which have already been practised.'

(Paragraph 321)

Relevance of the maths for our students should certainly be an important aspect of our teaching but we must beware of regarding it too superficially and as a general panacea. A relevant context will be an *appropriate* context and for some students this may mean an appeal to imagination/discovery/artistry, whereas for others it may imply *familiar* everyday contexts or application in other subjects being studied. Whatever the context the students' knowledge of maths must be sound. As teachers we need to be aware both of the level of difficulty at which we are 'pitching' the maths and the level of familiarity/relevance that the context presents.

Successful problem solving may therefore represent the ultimate in understanding. The ability to apply concepts and skills in many and varied contexts, transfer of learning, is the educator's vision. It is our task as teachers to help our students in the gradual building-processes that lead to such mastery... but how can we do this?

Making maths relevant is one aspect of our teaching but not all students are happy with problems or puzzles or even pastimes. And problem solving situations can cause anxiety and affect attitudes to study, so there are many factors to consider. We shall do our best within the space of this chapter to discuss some of the work we have carried out on the effects of context, attitude and anxiety on students' performance in maths.

Context

In our early work we found that embedding the mathematics in context appeared to create some tasks that had core of difficulty characteristics and some that had not. What then was the difference between these two categories of tasks? Was it the nature of the mathematics embedded or was it the effect of having a real problem to solve?

We talked to craft students about context:

'Well, it motivates us to learn... makes it real... and its OK in class. But when we go away to do it ourselves its difficult to sort out the calculations from all the junk... don't know where to start.'

Looking

Let's look at one of these early computational tasks involving a whole number and a decimal fraction. The task was set in an algebraic form and also within the practical context of Ohm's Law. This meant that it was relevant to craft students who were studying electricity as part of their course. Later, because of the results obtained from this further-education study, we repeated the exercise with school students and others, most of whom would have been exposed to Ohm's Law at some stage during their school career. Even if this was not so, you will notice that the task was presented in such a way that they did not need to have been particularly acquainted with the context. This background information is important to have in mind when you study Table 10.1.

Table 10.1

Task 1. Pure form: $8 \div 0.16$
        2. Algebraic form: if $0.16x = 8$, then x is?
        3. Ohm's Law context: given that resistance
          $= \dfrac{\text{potential difference}}{\text{current}}$

an appliance carrying a current of 0.16 amperes with a potential difference of 8 volts has a resistance in ohms of?

	% choosing					
	(1)		(2)		(3)	
Students	Corr	Incorr	Corr	Incorr	Corr	Incorr
	50	2	50	None of these	50	½
14/15+ CSE	25 ⎫ 59	—	28 ⎫ 59		24 ⎫ 60	
14/15+ 'O'	76 ⎭		74 ⎭		75 ⎭	
Craft 17+	19 ⎫ 38	48	23 ⎫ 44	31	29 ⎫ 45	26
Technician 17+	59 ⎭	—	65 ⎭		65 ⎭	
Teacher trainees						
Primary	70 ⎫		52 ⎫		56 ⎫	
Secondary	70 ⎬ 71		71 ⎬ 69		73 ⎬ 71	
Further-education inservice	75 ⎭		82 ⎭		83 ⎭	
Undergraduate engineers	91		98		99	

Corr, Correct; Incorr, incorrect.

In previous tables in this book we have, because of space, combined the performance of the 'O' level and CSE level students, the craft and technician students and the trainees and inservice teachers. Misconcepts associated with the core of difficulty exist for all learners but to a greater extent with CSE, craft students, and primary teacher trainees. However, when context is included the situation is more complex and therefore in this table we have listed these groups separately. If you study Table 10.1 carefully you will probably agree that the context

- Helped:
  craft and technician students especially the craft; engineers (although these students did well on all three forms) and the further-education business-studies lecturers.
- Made little difference to:
  CSE or 'O' level students in their fourth or fifth year at school (these students, on the whole, were meeting more formal mathematics) and the secondary-school trainees, many of whom were preparing to teach craft, design and technology in schools.
- Was disadvantageous:
  to the primary-school trainees; note that for these students the sight of an 'x' frightened them even more than Ohm's Law!

It would appear then that the context helped most of those students for whom it was relevant or who were old enough to have experienced in their work a variety of contexts as with the further-education lecturers. But what about the effect of the mathematics? The task **8 ÷ 0.16** is not quite a core of difficulty task: in the spectrum of difficulty it lies between the mechanical (both whole numbers) and the diagnostic (both decimal fractions).

In another task easy whole numbers were used but the information was presented with the aid of a circuit diagram as shown.

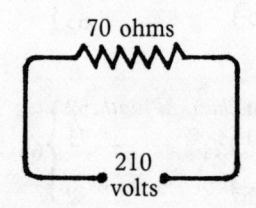

70 ohms

210 volts

Results were compared with the previous context task. It was found that the success rate was even higher for the same students but even more disadvantageous for primary trainees and this time for 'O' level school students.

In this problem the computation was mechanical but the context was too technical and hence neither familiar nor relevant for certain kinds of students. This problem thus highlights the importance of a context appropriate to the learner. Let's look more closely at these two components of problem solving. Broadly speaking there are two categories of mathematical content: the mechanical and the inferential (diagnostic). There are many tasks which fall between these two as with the first one discussed but at least these categories help us decide whether a task is more of one than the other. The context itself needs to be examined carefully because it involves processes associated with problem solving in general. Context also could be classified as 'mechanical' (e.g. familiar) or 'inferential' in the sense that it is too technical/irrelevant/unfamiliar.

There seem to be at least four categories into which problems can fall. We shall apply the labels mechanical and inferential to context, they may not be the most suitable, but they do imply a generality that is needed for the discussion.

● Mechanical mathematics embedded in a mechanical, i.e. clearly defined, context.
● Mechanical mathematics embedded in a more inferential context, i.e. the learner has to infer not only what to do but why and when.
● Inferential mathematics in a mechanical context.
● Inferential mathematics in an inferential context... the most difficult of all!

Thinking in these broad terms should also give us a basis to think more clearly of those situations where both the mathematics and the context fall between the mechanical and the inferential. We hope that the examples selected and the discussion which follows, though necessarily brief, will help you in your own thinking of problem solving.

Mechanical context and mechanical mathematics

Task   Pure: the average of 15, 17, 18.25, 15.75 is?
       Context: the weekly wages of four teenagers are £15, £17, £18.25 and £15.75; the weekly wage is?

Pure | 49% correct | 24% correct | 56% correct |

Context

54% correct	31% correct	70% correct
13/14-year-olds (mixed ability)	13/14-year-olds (slower learners)	TOPS trainees (19–57 years)

Although we cannot generalize from these results of course they certainly do suggest that a straightforward familiar everyday context does help the younger learner and the disadvantaged older learner.

Mechanical context, more demanding mathematics ($\pm$ common fractions)

Task Pure: $1 - (\frac{1}{3} + \frac{1}{4} + \frac{1}{8})$
Context: a recent government report announced that the 'average' family (a couple with two children) spends one third of its income on housing, one quarter on food and one eighth on heating; the fraction left for everything else is?

Pure	27% correct 23%...$\frac{3}{15}$	9% correct 48%...$\frac{3}{15}$	56% correct 18%...$\frac{3}{15}$	36% correct 14%...$\frac{17}{24}$
Context	40% correct 20%...$\frac{17}{24}$	26% correct 34%...$\frac{3}{15}$	60% correct 16%...$\frac{3}{15}$	33% correct 18%...$\frac{17}{24}$
	13/14-year-olds (mixed ability)	13/14-year-olds (slower learners)	14/15+-year-olds ('O' & CSE)	TOPS trainees (19–57 years)

The incorrect response $\frac{3}{15}$ is the usual trick of 'adding tops and bottoms' and this response was common to the school students. The incorrect response $\frac{17}{24}$ is the result of correct addition of the fractions but not subtracting this from 1 (note that this was chosen by TOPS students). A close study of these results
● Reinforces the importance of the mathematical content in determining success rates.
● Strongly suggests that a straightforward familiar real-life context helps younger learners and slower learners. Not only is the success rate higher but their incorrect choice of $\frac{17}{24}$ became a 'better' choice than $\frac{3}{15}$.

If it is possible to summarize, the results suggest that straightforward contexts which are relevant in some way (not necessarily from real life), or familiar, do help younger students and the slower learners to be more successful. Having said this it must be pointed out that real-life contexts have their own practical constraints as we pointed out in the examples with percentages. If the context relates to eating out, then while a student may be able to calculate 10%, what is actually given for service does not depend on mathematical thinking entirely. This is why it is so important, when discussing the solutions of mathematical problems with our students, to include the relation to commonsense and real-life activities.

Inferential context, what do you think of the maths?

Compare these two tasks. The mathematical content is not quite the same... but does it matter?

Task Pure: 0.000011 x 100
Context: for an increase in temperature of 1°C a 1mm length of steel will expand 0.000011 mm. For an increase in temperature of 5°C, a 20 mm length of steel will expand by?

	CSE 14/15+ 'O'			Craft 17+ Technician		
Pure	60% correct	75%	84% correct	45% correct	60%	77% correct
Context	20% correct	42%	48% correct	24% correct	34%	44% correct

	Primary Teacher trainees		Further-education inservice	Under-graduate engineers
Pure	78% correct	84% correct	93% correct	98% correct
Context	41% correct	57% correct	71% correct	79% correct

The contrast in success rate for all groups is quite dramatic. The context task was a 'core of difficulty' task. The most popular incorrect response selected was 0.000055 (0.000011 x 5); it was the choice of 35% of CSE students and 36% of craft students. The mathematics of this task was 0.000011 x 20 x 5 ... was this causing the trouble? Or was it the context, which would be familiar to students who had studied some physics? Is 0.000011 x 20 x 5 really much more difficult than 0.000011 x 100? We presented this in both pure and context forms to CSE and 'O' level students and found that the success rate for 0.000011 x 20 x 5 was considerably less than for 0.000011 x 100. Listening-in to some students solving the task went typically like this

'0.000011 times 20 is... times 5 is... '.

and doing it this way of course easily leads to confusion and error. They did not scan the task as a 'whole', spotting that 20 x 5 is 100. Their solution was mechanical, following the order given and not appreciating that the order of multiplication does not matter.

Inferential context, inferential maths

It is clear that this category would provide the most difficult problems of all and you could no doubt provide many of these for your students. What is clear is that when we design problems or select them we need to take care over both the context and the mathematics involved.

The 'diagnosis' of context may be even more subtle than that for mathematics. Opportunities for mis-match between teacher and learner are plentiful! What may be familiar context for the teacher may well be completely outside the experience of the learner. The teacher judges the context to be mechanical but for the learner it may be inferential because it is neither familiar nor relevant. Recent analyses of the examination results for some thousands of 17+ students on tasks involving mechanical maths skills in everyday contexts have provided us with this kind of evidence.

For the classroom, there is a gradually increasing supply of resource material with examples from real life and the world of work and also appealing to the imagination. A context *appropriate* for our students is what matters : young learners are not necessarily interested in problems involving mortgage rates!

Perhaps a final example would emphasize the importance of context.

Task A twist drill running at 600 rev/min has a feed rate of 0.15 mm/rev. The feed rate in mm per second is?

Rev/s is $\frac{600}{60}$ (or 10)

Feed rate (mm/s) is $\frac{600}{60} \times 0.15$

A wide range of students attempted this problem and the most popular response was 0.015, that is they worked out 10 rev/s and then *divided* 0.15 by 10. So some students multiplied, others divided. The computation was reasonably straightforward but knowing whether to multiply or divide depended on understanding the context.

...twist drill?... *How relevant can you get?* *mm...per sec... ...must be smaller... ...÷...*

This kind of situation is paralleled in the CSMS work where younger learners asked 'Please, miss, is it an add or a multiply?'
Summing up then: when we design problems for our students we should ask ourselves

- Is the mathematics involved 'mechanical' or inferential/diagnostic, i.e. does it involve tasks with core of difficulty characteristics?
- Is the context appropriate, is it straightforward or inferential in its demands?

As long as we know what it is we are requiring of our students then we can mix and match between the categories of problems... always being *aware*, *explicit* and *flexible*.

Attitude

'The shutters appear to come down whenever the word mathematics is mentioned'. This was one of the first statements made by a further-education college lecturer about his students in the early days of our studies. It expressed a sentiment which has been echoed through succeeding years

years concerning may different kinds of students from school to adults on retraining courses. Possibly one of the most alarming reactions to a brief list of questions concerning attitude to mathematics came from the primary-school teacher trainees. Some of these students volunteered written additional comments such as 'I *dislike* mathematics: I never could do it at school... I'm frightened of having to teach number to young children'. Our children's attitude to mathematics may be set for life at primary level... is it possible that not only is there a learning–teaching–learning cycle with respect to common misconcepts but also to attitude towards mathematics?

Yet this rather depressing finding is of course only one aspect of the work at primary level: there is also much richness and excitement. This was expressed aptly by several 10-year-olds in response to the question 'Do you like maths?', who said:

'I think it's brilliant, it's nice, good fun doing it... .'
'I think it's great... .'

And in contrast:

'I sometimes think it's a necessary evil...' !

The usefulness of maths

Parents' attitudes to maths may well influence those of their children. In a desire to show sympathy their own feelings of incompetence may be passed on. Nevertheless parents seem to have a positive effect in conveying to their offspring the importance of learning maths 'because it's useful'. Indeed indoctrination seems to start at an early age.

Here are some of the observations made by 10–12-year-olds with whom we discussed the usefulness of the subject.

'Well I think if you're working in a shop you have to add up numbers... if someone's going to buy something... and if you're working down at the airport you got to still know how you add up... and if you work at anything else you still got to know how to add up...'.
(And following on from a discussion of aircraft)'... yes you've got to, you should know what times they come in and what times they go out, so if people phone up and ask what time our plane's coming up you can tell them then and there'.

It was disappointing to find younger and older students at school very limited in their appreciation of the uses of maths: it did not seem to extend far beyond the boundaries of the supermarket. The notion of it being a concise and powerful form of communication, its power to generalize thinking, was, sadly, absent. There were, however, some novel uses such as this:

'Interests, hobbies... maths seems to be all around... swimming (how many strokes); sprinting (how many lengths from the end should you start sprinting?)'.

And this discussion indicates the influence that a parent may have, especially Mum!

I: Do you think your Dad finds maths useful?
S: Yeah, he might.
I: Do you think your Mum finds it useful?
S: Oh, yes, *very* useful.
I: Does she talk to you about it?
S: Yes, she tells me that she's got to divide say 78 into 2, things like that.
I: When does she have to do that sort of thing?
S: When she's doing typing, because when she has to get it centred or something like that she has to do these sort of divisions and adds and that, to find out where the centre of the page is... .

Craft and technician students were clearly in a position to see the relevance of maths in their training. Many regretted their lack of understanding and competence when at school and these regrets were causing a lack of confidence in their vocational studies. The perception of most people regarding the importance of and the usefulness of a knowledge of at least some mathematics is reasonably well documented. Although it is lamentable that this perception is rather limited, nevertheless such an attitude by the community should be used to advantage in our teaching. There is no such parallel with *liking* mathematics. Feelings range from intense dislike through indifference to liking and enjoyment of the subject.

Liking and performance

We found, as might be expected, correlation between teacher trainees' liking of the subject and their performance. The surveys at primary and secondary levels carried out by the Assessment of Performance Unit (APU) have included studies of attitudes and these have shown correlation between school students' liking and performance. The second international study of mathematical achievement of 13- and 18-year-olds has recently been conducted and has included a study on attitudes. The National Foundation for Educational Research will soon publish the report on the UK findings.

Dislike and anxiety

As maths teachers many of us have probably experienced the dislike and even fear which our subject can generate, not only among friends and acquaintances, but with our professional colleagues.

> S.D.? Help! Deviations in maths? ...standard?

> $\pm \ \times$
> $\pm 1.7 \quad 0.1 \times 0.1$
> Oh dear - back to school!

Adults' panic about maths and the blocking of their reasoning by emotion is fully discussed by Laurie Buxton in his reporting of several case studies. Some saw maths as having a sort of mystique: 'I dislike doctors because they seem to have wrapped themselves round in magic – mystique – and I feel mathematicians have the same quality.' Did he see mathematicians as priests? 'Yes' And do they have power? 'Priests always have power, don't they?'.

Such strong feelings can create deep anxiety for adults and children. Putting maths into contexts, even everyday contexts, does not necessarily reduce anxiety... it may even increase it as is implied by our previous discussion. Our own studies, with many different students, in the language laboratory revealed on occasion feelings of utter frustration and even panic; mostly such feelings were generated by the core of difficulty tasks. Performance, attitude and anxiety therefore seemed to be inextricably bound. We felt there was a need to investigate this interaction... and where better to start than at the primary school?

Anxiety, attitude and performance

Students' attitudes may already have been set by the time they reach secondary school. It therefore seemed sensible to carry out a study on the interactions between attitude, anxiety and performance as early as possible.

One of our objectives was to study whether the two kinds of mathematics, the mechanical and the inferential (core of difficulty) also existed for younger learners. We therefore decided to work with children in their final year of primary education. This allowed us to use the paper and pencil techniques and strategies of our earlier studies. It also enabled us to explore the transition from primary to secondary level.

Here is a brief summary of the work.

The study: looking, listening and learning

Looking

We wanted to find out, firstly, whether the two types of mathematical thinking appeared at these earlier stages of learning and, secondly, to examine how negative attitudes and anxieties affected performance. Selecting and designing tests at primary level is rather difficult because of the variety that exists across schools with syllabus, materials and teaching methods. Much of the material available is also not suitable for these early stages and so the tests used had to be modified or designed from first principles.

The mathematics test was designed on the basis of what we had learnt from tests given to older students. Analogues of many of these tasks had to be devised. A novel feature of this test was that each test included two short 'I think' type questions which required these younger students to make an estimate of the difficulty of the task and their chance of success or failure *before attempting it*. Detailed verbal instructions were given on how to answer the 'I think' questions by putting a point on a given line.
Here is an example:

Task What number must go in the box to make the following true:
$\frac{2}{\Box} = \frac{1}{3}$

I think that this problem is?

        Easy_____ difficult

I think my answer to this problem will be...

        correct_____ wrong

ANSWER:_____

These two scales were intended to provide us with information on two points. First, what would be the relation between these younger *students'* assessment of their own difficulty and their actual performance; would there be a disparity? (We already knew that for the older students there was a large disparity between their actual performance and their *teachers'* assessment of difficulty). Secondly, compounding of the two ratings on the two scales, for each task, was designed to provide a measure of self-confidence. A student who judged that a problem was difficult, but that she/he would nevertheless get it right, was awarded a high score while the combination of 'easy' and 'wrong' led to a low score.

We were also concerned with the effect of anxiety on performance, concentrating on two kinds of anxiety: anxiety which is specifically task-related (test anxiety) and anxiety which is a characteristic of personality. The anxious personality tends to lack self-esteem and self-confidence; would this be reflected in performance? Other more explicit measures of anxiety were used as were measures of attitude and of the structure of personality. How did all these relate to the two kinds of maths tasks?

Listening

As with the older students we combined the *looking* at written work with the *listening* to thinking aloud. A selection of maths tasks was made on the basis that they had presented unexpected difficulties or revealed interesting characteristics during the group testing. These tasks were then talked through on a one-to-one basis with an interviewer. Here are two tasks you've met before: the same 9-year-old was involved.

Task **What number must go in the box to make the following true?**
$$\frac{2}{\Box} = \frac{1}{3}$$
I: Have you seen this type of sum before?
S: Yes... its six.
I: Why?
S: 'Cause that... two is double one... so six is double three.

Task **What number must be placed in the box to make the following true?**
$$\frac{\Box}{8} = \frac{3}{12}$$
S: 'Well... eight goes into... that... well... (laughs)... sorry... you times that by... you've added four to that... oh!... hold it I'm stuck!... ah...

(laughs)... right! so... that... you've added four, cause eight goes... four and eight... I mean four goes into eight and twelve so... so... so... it goes... mmmm... so four goes three times into that... ooooooh!
I: You're doing well. Keep trying.
S: So it goes two into eight so it I think it will be two.

Do you notice the difference in the way these two tasks, apparently similar, were solved? What did we learn from this kind of listening and from the looking?

Learning

● Two types of mathematical tasks were identified which related to those discussed for the older student. The 'inferential' type of thinking, however, was in a much more emergent state at primary level and became more evident at the early secondary level. This mode of thinking thus appears to be developmental... it may well parallel the development of the formal reasoning proposed by Piaget?
The distinction between the two types of task was confirmed by the interview studies. The algorithmic tasks trigger rapid solutions via well-learned techniques and the inference tasks generate characteristic 'fumbling' toward a solution involving restructuring of the task.
● Primary students were generally capable of a fairly realistic assessment of difficulty for the more 'algorithmic' tasks but were unable to do this for the 'inferential' tasks.
● For secondary students (but not for primary) self-confidence scores correlated positively with success at solving inference tasks but not at all with the other tasks.
And what did we learn about the effects of attitude and anxiety on performance? It is not simple, as one might suspect.
● If anxiety is measured as a stable personality characteristic there is no difference in anxiety levels between late primary and early secondary students.
● For specifically test-related anxiety, however, the level for primary students is greater than that for secondary.
● For both groups a high level of test anxiety is related to poorer mathematical performance. It follows that the disadvantageous effect of anxiety on performance is of greater importance at primary level.

- For primary students the extent to which anxiety impairs performance is dependent on the *attitude* to maths.

For those who have a positive attitude, high levels of anxiety have less effect than for those with negative attitudes. There is no evidence of this relation for the secondary students.

- There is no evidence of any marked differences in attitudes to maths between the primary and secondary students. However, with regard to attitude to school generally the primary students had more positive attitudes and regarded school as a more useful and friendly place than did the secondary students.

Thus it would seem that the relation between attitude, anxiety and performance is a much more complex and potent factor at the late primary level than at the early secondary level. It may be that the relation is masked by other factors which are much more dominant at the secondary level, changes in methods of teaching, content, school organization and so on? After all, 2 years on in the life of a 10-year-old is a very long time.

How then can we use these findings to our advantage and that of our students? To what extent, for example, can we reduce anxiety to a level at which it becomes an incentive to learning? What can we do for the students with a strong tendency to anxiety in test situations?

Prescription and aftercare

Aim for ourselves AWARE EXPLICIT and FLEXIBLE
    for students CONCEPTS COMPETENCE and CONFIDENCE
                           ... TRANSFERABILITY

Here are some brief suggestions for you to think about, discuss with your colleagues and possibly put into action.

Suggestions

Reduce test-related anxiety

Spend the time and effort on attempts to prevent the development of test-related anxiety. What about a programme of regular and frequent testing designed with these students in mind? There is nothing like success to breed success. Each test should therefore be at a level of difficulty low enough to ensure success and yet containing enough of the 'failure' material to provide the opportunity for *supportive additional help* and the retrieval of failure.

Transform the inferential to the algorithmic

For all our students use our ability to distinguish between the mechanical/algorithmic and the inferential tasks, between those that don't and those that do really unlock the blockage to progress. Even the inferential tasks are amenable, at any age, to teaching via the use of standard algorithms, i.e. it is possible to convert all tasks into an algorithmic type by appropriate teaching techniques. But let us ensure that our students' use of these algorithms is based, as far as is realistic, *on a sound conceptual grasp of the material*. And this of course assumes that our expectations as teachers must be in line with our students' capacities.

Generate positive attitudes, confidence, flexibility... and transferability

- We need first to generate *positive* attitudes. This may be done through puzzles and pastimes, investigations and other creative activities which may not of themselves achieve learning but they will provide the springboard for learning to take place.
- Build up our students' *confidence* through a repertoire of successful routines built on understanding. These are the essential sharp instruments of problem solving.
- Implant these routines in a few familiar/relevant and straightforward contexts; increase the number of contexts; generalize to a wide variety of contexts.

And all this will involve a widening of the 'curriculum': the role of maths, the history of maths, maths across the curriculum, multicultural maths, *talking* maths, fears about maths, worries and enjoyment. Broadening our students' experience, building up their confidence; isn't this what schooling is all about? And this broadening should be available to *all* our students, slower learners as well as faster learners, so that all have at least some appreciation of the power of mathematics to communicate. Only then may the educators', and now, the politicians' dream come true... transferability.

Some questions for discussion

1. What do you see as the role of context in your own teaching of maths? How do your colleagues see it? Discuss.

2. Does context motivate your students? Does it help them with their understanding of the maths concepts? Does it help them better understand and become more skilful with the techniques?
3. Have you tried different kinds of contexts in your teaching? Do you find that students of the same age respond in the same way to different contexts? Do you find that students require different kinds of contexts depending on their ages?
4. Does solving a problem (maths in context) create more or less anxiety for your students (all, some?) than a straightforward tackling of the 'pure' maths involved?
5. Have you any observations of your own students on the interaction between anxiety and attitude in problem solving?
6. What aims do you have in mind when you construct tests for your students? What factors do you take into account?

CHAPTER 11

ADULTS AND NUMERACY

'You didn't make anyone feel like a dummy because they were experiencing difficulty with learning the subject'.

TOPS trainee

The Cockcroft Report *Mathematics Counts* makes the observation:

'There are, of course, many people who are able to cope confidently and competently with any situation which they may meet in the course of their everday life which requires them to make use of mathematics. However, there are many others of whom quite the reverse is true.'

(Paragraph 19)

The people referred to here are adults.

Concern about adult *literacy* can be traced back to the early/mid-1970s with such landmarks as the BBC programme *On the Move* and the establishment in 1975 of the Adult Literacy Resources Agency which was set up to help LEAs and voluntary organizations to provide adult literacy provision to complement the BBC initiatives. It was from this early work on literacy that it became apparent that there was an equally important adult numeracy problem. This growing awareness has a similar history to that of adult literacy with television landmarks of *Make it Count* (ITV) and *It Figures* (BBC).

Diagnosis

One of the striking features of the adult numeracy problem is as the Cockcroft Report states:

'the extent to which the need to undertake even an apparently simple and straightforward piece of mathematics could induce feelings of anxiety, helplessness, fear and even guilt... .'

(Paragraph 20)

People have said to us:

'I hate maths.'
'I avoid numbers if I can help it.'
'I missed my first-year maths classes and never caught up.'
'Couldn't see the point of algebra.'
'I had a rotten maths teacher and so drifted to the bottom of the class.'

And what about this one:

'Maths makes me feel sick.'

School leavers are tomorrow's adults. We have already seen in earlier chapters that what we learn at school matters. The types of error made by the adult are often the same as those made by the 13-year-old. However, the difference between the adult and the school leaver is that the school leaver may find it easier to relearn the numeracy that is needed, whereas the adult may have such deeply rooted misconcepts that more time and effort is needed for the relearning process.

The fears and anxiety that a school leaver may have about numeracy will probably fade as they leave the academic world behind. They will learn to cope with the mathematical demands of life as and when they arise with a 'muddle through' attitude. However, their problem will rise up in front of them periodically when they are confronted with a clear mathematical situation that they cannot tackle. We need to realize the depth of negative feeling to be able to help our students.

Curriculum initiatives have taken place to help solve the problem. For example, in the early seventies the City and Guilds of London Institute (CGLI) received requests from schools to provide schemes for young people in their final year at school. As a result of a number of investigations and contact with representatives from education and industry the CGLI developed a number of vocationally oriented schemes (foundation courses) that would aim to motivate young people to improve their basic skills of literacy and numeracy by relating these to the world of work.

In addition to the need expressed by the schools many colleges of further education showed concern about the difficulty experienced by students with the mathematical content of their first-year craft studies. Our research into the mathematical achievement of school leavers confirmed a lack of basic numeracy skills and also revealed that these students experienced a real sense

of failure in mathematics with consequent anxiety and fear about their college course and the effect on their job. The CGLI numeracy (364) scheme was consequently developed with the aims to:
- Raise their standard of numeracy to a minimum level required for day-to-day living and working.
- Gain confidence in their own ability to learn and to manipulate numbers.
- Cope with further work in calculations required for present and future study.
- Transfer number skills learned into new situations.

Adult and for that matter 17+ trainees often request help on a variety of topics concerned with basic numeracy. The first challenge for them is to acquire the confidence to master their difficulties and at the same time begin to overcome any fear of mathematics or number that they may have. Our job is to help and guide them so that they are able to acquire and use the basic concepts and skills that they need for everyday life and work.

The way we teach adults and 17+ students will be different from the way we teach younger students. We need to remember that these older students have been 'through the mill' before; that they may have semi-consistent concepts which will need building on, maybe destroying. We need to be aware that these students may lack concentration and be tired because they have worked all day. Some students have difficulty remembering specific facts; a problem that can develop with age. They often find it difficult to learn new methods since their own old erroneous methods are so deeply ingrained in their memory. On the other hand, it is important to realize that a student may have knowledge and expertise in areas of which we as teachers know very little. This can be very useful since it gives an opportunity for all to learn something. We should use their 'life experiences' to motivate them in ways that are not always possible with younger students. If we give these students 'more of the same' we will simply reinforce their negative attitudes; a new approach must be sought.

The requirements of adults are not just varied, they are often very explicit. For example, 'I need to pass the Job Centre arithmetic test to get on a training course', 'I don't understand how my mortgage works' and so on.

We should aim to teach basic numeracy skills to adults so that they can both 'transfer' the mathematical requirements of a problem into a task and confidently tackle the task. This means exposing them to different contexts which have the same mathematical content, with the aim of developing flexibility in the student.

The first few meetings

Looking

The environment which you as a tutor work in is not always what you desire; make the best of it. Decisions need to be made that will dictate the type of atmosphere that you and the students have to work in. Desks in rows, for example, tend to generate a formal atmosphere, while desks clustered in small groups or in a 'u' shaped formation create an informal atmosphere. Walls covered with bright posters or charts are warmer than walls that are bare.

Listening

At your first meeting with your students discuss their feelings about numeracy. Let your students talk. It is often possible to observe common points concerning feelings of strengths but more frequently the discussion will be biased toward weaknesses, negative attitudes and anxiety. Sometimes comments are made on where things went wrong. For such a discussion session to be successful it is essential to generate a friendly and relaxed atmosphere.

Learning

During this first session it is important to pick up cues, take notes; this is all part of the diagnosis. Past experiences play an important part in the decision process on how and what is to be learned.

Diagnosis and assessment

An informal assessment may be less alarming for an adult. One method of diagnosis using an informal assessment is to construct a number of displays of 'life skills' materials (such as those on p. 197). The students could be given answer sheets which they fill in as they read each display. The displays could involve

- Interpretation exercises. For example, read the newspaper article on unemployment: how many people/what percentage are unemployed?; planning a journey; how much will it cost; how long will it take? shopping: which is cheapest? and so on.
- Practical exercises, e.g.: weighing yourself; writing down your weight; taking your waist measurement; measure a window to fit a curtain.

● Pure calculations, e.g.: change from a £10 note if your bill is £6.75; the cost of half a pound of tomatoes at 45p a pound; temperature drops five degrees from −2°C, what is the temperature?

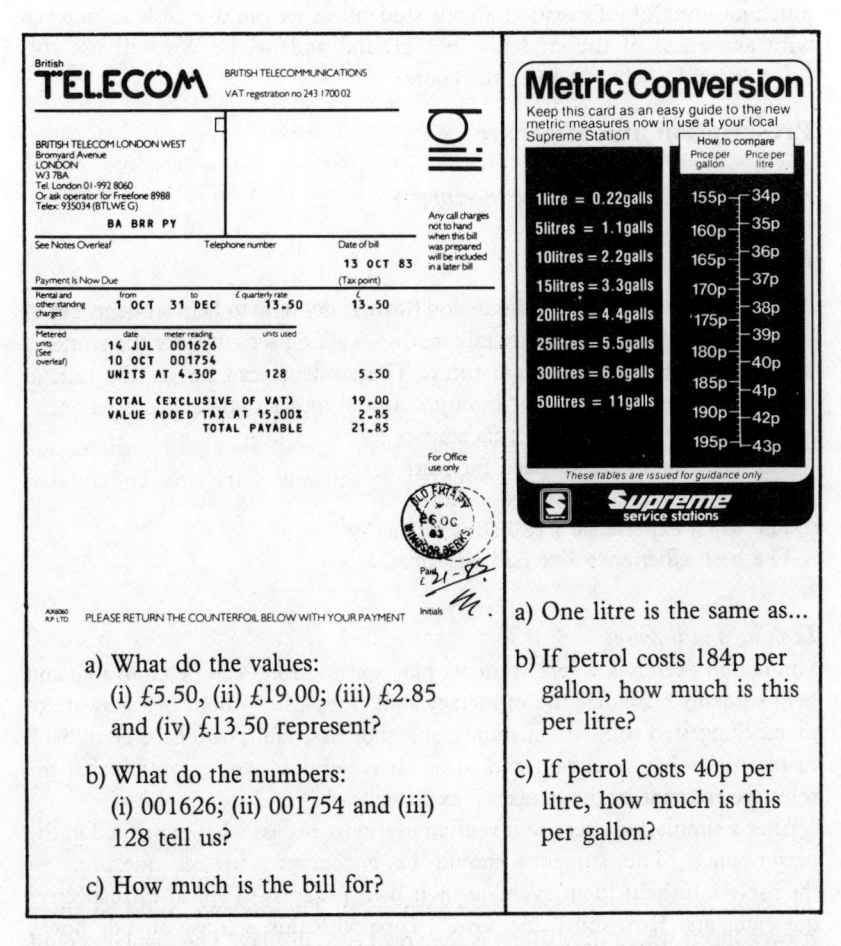

British TELECOM
BRITISH TELECOMMUNICATIONS
VAT registration no 243 1700 02

BRITISH TELECOM LONDON WEST
Bromyard Avenue
LONDON
W3 7BA
Tel. London 01-992 8060
Or ask operator for Freefone 8988
Telex: 935034 (BTLW E G)

BA BRR PY

See Notes Overleaf | Telephone number | Date of bill
 | | 13 OCT 83

Any call charges not to hand when this bill was prepared will be included in a later bill

Payment Is Now Due (Tax point)

Rental and other standing charges: from 1 OCT to 31 DEC £ quarterly rate 13.50 £ 13.50

Metered units (See overleaf): date 14 JUL meter reading 001626
10 OCT 001754 units used
UNITS AT 4.30P 128 5.50

TOTAL (EXCLUSIVE OF VAT) 19.00
VALUE ADDED TAX AT 15.00% 2.85
TOTAL PAYABLE 21.85

For Office use only

AX6060 R.P. LTD PLEASE RETURN THE COUNTERFOIL BELOW WITH YOUR PAYMENT Initials

a) What do the values:
(i) £5.50, (ii) £19.00; (iii) £2.85 and (iv) £13.50 represent?

b) What do the numbers:
(i) 001626; (ii) 001754 and (iii) 128 tell us?

c) How much is the bill for?

Metric Conversion
Keep this card as an easy guide to the new metric measures now in use at your local Supreme Station

How to compare
Price per gallon	Price per litre
155p	34p
160p	35p
165p	36p
170p	37p
175p	38p
180p	39p
	40p
185p	41p
190p	42p
195p	43p

1litre = 0.22galls
5litres = 1.1galls
10litres = 2.2galls
15litres = 3.3galls
20litres = 4.4galls
25litres = 5.5galls
30litres = 6.6galls
50litres = 11galls

These tables are issued for guidance only

Supreme service stations

a) One litre is the same as...

b) If petrol costs 184p per gallon, how much is this per litre?

c) If petrol costs 40p per litre, how much is this per gallon?

A formal diagnostic assessment, similar to the one at the end of the book could also be useful. It will provide an overview of the students' ability in the topics covered in the assessment. The anxiety aspect of setting a formal written assessment, however, needs to be considered carefully for many adults; a written assessment may resurrect forgotten bad memories of school maths which may result in underachievement.

We should be looking for strengths and types of errors made during the course of the assessment; gathering information which we can analyse and interpret to help us determine a good starting point. We need to collect as much meaningful information on our students as we can to enable us to teach with awareness of the students' background and needs. We will use this information to help structure the course.

Prescription and aftercare

Activities after the first few meetings

Listening

At later meetings encourage discussion further; not only to help students get to know one another but more to 'talk maths'. Talking activities for the students are good listening times for a tutor. The students can be set the task of interviewing one another, for example. Initial topics that can be used are
  What I want to get out of this course
  The maths/numeracy I need for work
  The maths/numeracy I need at home
  The worst experience I've had with maths
  The best experience I've had with maths.

Looking and listening

Simulation exercises where students play specific roles can be enjoyable and help students recognize the numeracy skills they use without being aware of them. Suggested roles are customer and shop assistant/post office clerk/bank cashier/ticket-box attendant and so on. It is helpful, even essential, that the roles are relevant to the students' experience.

After a simulation exercise it is often useful to discuss what happened in the performance. The students should be encouraged to ask questions of themselves to help them evaluate their own progress. Typical introspective questions are: How did I do?, Can I cope with this exercise outside now?, How did I feel about doing the exercise? This kind of activity will help the students determine what they have learned and what still needs to be learned or practised further.

Other simulation exercises which do not require the students to play a role but are 'life skills' related include preparation of a day trip, decorating a room, dieting, cookery and budgeting.

Learning

Ideally we should aim to teach basic skills through the student's own particular situation, so that the skills would be taught in a context relevant to the student (but in the light of the discussion in Chapter 10). However, we are all aware that number work must be well practised, especially if we want speed as well as accuracy. It is essential that we encourage our students to develop their newly learned skills outside the sessions they are together. It is equally important that we also revive previously learned skills otherwise they will be forgotten quickly. To this end we should try to expose our students to as many different contexts as they can cope with once they have grasped the basic idea.

Lesson plans

We are all aware of the need to plan lessons in advance. By doing so we know what we intend to teach and will have prepared any material before we begin our session. However, our lesson plans must be flexible to allow for deviations from the plan whenever the need arises. It is good practice to have a lesson structure that recaps what was covered in the last session followed by an introduction to the work to be covered in the current lesson. A pattern of 20 minutes talk and explanation followed by 20 – 30 minutes of student application of the ideas developed will allow close monitoring of student progress.

During the session we need to keep in mind the
Aims of the session
Student needs
Information collected from our diagnosis
Materials available
The success of the session.

Recording and evaluating progress

It is clearly important to record what a student has covered and how successful the student has been. Records can be kept in a number of ways, although it pays to aim for simplicity.

The records should include brief notes on what has been taught, the problems encountered and suggestions for future activities. We should encourage our students to keep their own record of their progress. This can be

done in a number of ways. For example, you could:
- List the objectives set for the session and ask the students to rate their confidence on these objectives using a three-point scale as Confident, Not so confident, Can't cope.
- Construct an open-ended questionnaire with questions like: What work have you covered this session?, What can you do now that you couldn't do before?, What do you still have problems with?, What do you want to work on next?
- Ask the students to write about what they have covered and how they feel they are progressing.

The principle of making students think about and plot their own progress may appear unusual but it can make them more aware and effective learners. We must supply positive experience to build up their confidence to help break down negative feelings toward the subject. When the student asks the question 'How did I get on today?' We want to hear a positive response. If progress is discussed within the class group the students will be able to share hopes, fears and doubts. Such group discussions allow the opportunity for individuals to gain insights into other people's capabilities as well as their own.

Remember we are aiming at
CONCEPTS COMPETENCE and CONFIDENCE in our students
while we must be
AWARE EXPLICIT and FLEXIBLE in our teaching.

Some questions for discussion

1. If a student gives an explanation of a solution that is different from your own, how do you react? Is your method the best? Is the student's method inefficient? If so, is it necessary to change it?
2. Incorrect answers are sometimes given very confidently. What implications does this have when you are trying to build up confidence? Consider the situation: VAT at 15% still has to be added to a bill of £20, how much VAT is added? Answer: 15p. How do you proceed from here?
3. Referring to the opening comment of the chapter 'you didn't make anyone feel like a dummy.' What implications does this comment have for your teaching?

CHAPTER 12

TEACHING MATHS

'As ten millions of circles can never make a square, so the united voice of myriads cannot lend the smallest foundation to falsehood.'

(Oliver Goldsmith)

Training for teaching

Education is full of paradox. Teacher training has its full share, nearly always brought about by examples of two conflicting human demands: for adventure and security. Sometimes we want one, sometimes the other; but we often want both and have to decide between irreconcilables. For ordinary school development we want free creative work with opportunities for experiment and trial-and-error learning and a feeling for real mathematical thinking; but we also want reasonable technical efficiency in frequently occurring situations and this is seldom acquired without hard work and practice. It is fairly clear that efficient learning (and enjoyable learning) is partly dependent on a repertory of both kinds of activity. Much the same holds for our own professional development. The needs of the preservice and the inservice teacher are slightly different but both need the right balance of adventure and security if they are to go on learning throughout their careers, and what a barren occupation teaching is if we fail to do that.

Preservice and probationary teachers need mostly to acquire confidence. Their ordinary training should begin to prepare them for presenting material at the right level for their classes; to keep their classroom language at an appropriate level; to work within some kind of conceptual map; to provide working material of the right kind of range and difficulty. Unfortunately, the 'difficulty' of a mathematical task is a complex notion. Attempts have been made to compare the difficulty of different tasks by assessing the number of steps involved in reaching an answer and by other rather mechanical means. These are useful in only a limited way. However, the beginning teacher partly measures success by how many of the class get 'right answers'. The most obvious warning coming from the diagnosis in these chapters is that it is all too easy to import extra and unnecessary difficulties into a lesson. New topics can be introduced while consciously avoiding some of the trouble spots identified in this book. This is not to suggest that we have to avoid all difficulties but more that those in the realm of the 'core of difficulty' will probably need much more in the way of specific preparation.

Established teachers will have the experience of marking their tests, exams and ordinary classwork and noting that, however well the current piece of work has been prepared and taught, the number of pitfalls that suddenly appear from past work remains daunting and can be discouraging. Anyone marking public-examination scripts is also well aware of the many errors; but, we suggest, not all. The cynic may ascribe them all to a version of original sin but many of the examples in the previous chapters show instances where imperfectly formed rules or an incomplete image in the mind works successfully for a range of problems but breaks down in critical places: the diagnostic ones which reveal the hidden difficulty. Nothing is more corrosive of childrens' confidence than to be continually reproached for 'carelessness' when what is happening is not careless but the elusive slipping of the anchor on which they have been relying.

The activities suggested here seldom require a major change of style (conventional to progressive, or vice versa) but do suggest feeding into normal working enough diagnostic material to check how well founded the concepts and skills really are. After that comes the step of arranging teaching programmes which deal with the particular difficulties explicitly and that may involve quite radical changes of pace and order in the curriculum. For example, work on fractions usually involves division of shapes into equal parts. At once there is a clash between the intuitive concept of area using notions of conservation demonstrated by 'equal to' and 'greater or less than' and notions of measure of area which are much more sophisticated. Teachers' discussions on these topics can get quite heated! The result may be a much more careful treatment of area over a longer period of development.

Diagnostic questions and diagnostic tests

In the activities that follow work with number and measure predominates but it is wrong to think of this as 'only arithmetic' or, even worse, 'only for the slow learner'. Ideas of sets, operations and relations underlie all the activities, whether made explicit or not. The concepts involved are all fundamentally mathematical ones and the images and models that students build in their minds are all examples of genuinely mathematical activity. The difficulties discussed are ones known to be very persistent so that they continue to appear in undergraduates, primary teachers in training and adults; not so much in high flyers or specialist mathematicians, of course, but some of their alternative problems are discussed under preservice training. An illuminating illustration for the authors comes when teaching statistics to Master's Degree students in social science. They are nearly all talented postexperience students in responsible jobs but they are just as vulnerable to these errors and misunderstandings, and consequently exposure to strong feelings of anxiety, as the children.

The diagnosis we are discussing is therefore different from some other uses of the word. It is not mainly about identifying slower learners. They will have far more problems than those discussed here. When working with slower learners we can put the results of a diagnostic enquiry to use because they need especially clear models and practices. So often our own are confused with hidden assumptions, jumps and short cuts about which we have stopped enquiring because custom and confidence have assured us of our 'right answers'. As always, our problem is to start where the learners are and to try to share their perceptions of the tasks in hand. When we are aware of the special difficulties ahead we can help by developing practices that will withstand the later pressures. How much damage has been done by old rules such as 'when multiplying by 10, just add a 0'! Another practice which will not stand up to pressure is allowing the verbal use of 'point sixteen' for '.16'; harmless enough in conversation but disastrous in calculation. This one is an example of difficulties built into the world we live in. It is so easy to say, 'You know how to do this sum with whole numbers; it is just the same in decimals' but it is not true, for the simple reason that we attach names with meaning to the digits of 163 (one hundred and sixty three) but that useful device does not exist for the fraction part of 163.27. The name of the number becomes a string of digits without labels.

Neither is our form of diagnosis strictly comparable with that of the common published diagnostic tests. The reason is merely that standardized tests are prepared for a different purpose. They are used almost entirely for classifying, placing the person tested in an appropriate percentile rank in a comparable group of people, usually an age band say of 11-year-olds. To do this certain technical criteria have to be met. The most serious for our purpose is that of 'predictive value'. For this, particular items have to withstand 'item analysis' in which items which are answered correctly by too many people or too few are regarded as not having enough predictive value and are therefore rejected. Out from these tests will therefore go some of the very items that we want to use for our purpose: which is to discover where people have their point of breakdown in the procedures they normally rely on.

There is a large literature on educational testing but beginners will find two publications of the Mathematical Association, namely *Tests* (1978) and *Interface* (1983), very helpful. The first gives a list of published tests with a brief guide to the technical jargon and the second discusses transfer from primary to secondary school with a chapter called *Tests* and other references to the problems we are dealing with. They are both referred to in the activities in this chapter. At a similar level *Mathematics Teaching Pamphlet* No. 14 (Examinations and assessment 1968) published by the Association of Teachers of Mathematics gives a much broader view of the whole question of assessment of mathematical development. It has now gone out of print but there may be copies languishing in libraries or stock cupboards.

Writing questions

The format of the questions in our tests is that of 'simple completion multiple choice'. This is now fairly familiar from public examinations but is often treated with suspicion because of fears about guessing and about the lack of sustained working out. However, arguments on these points can only be valid when they take into account the purpose of the test. No-one, presumably, feels that the multiple-choice format is a suitable way of assessing overall talent or ability but the main point about it is that it does some things extremely well. For busy teachers such tests have the real advantage of being very quick to mark, but they have the corresponding disadvantage of being extremely difficult to write. Subtleties in the wording and choice of alternatives cause quite considerable differences in performance and it is the experience of every writing team that a large fraction of the questions has to be rejected or modified. The questions in these tests have been exposed to considerable testing and have withstood the strains put on them by a wide range of students.

There are two slightly unusual features in our format. Four alternative answers are offered but a fifth alternative allows 'none of these is correct; I think the answer is... ' and occasionally a sixth is useful which allows 'I cannot do this one *or* I do not think we have covered this work'.

Terms used in the standard literature are:
Stem (for the question as posed)
Key (for the correct response)
Distractor (for the wrong responses).

We prefer to call the wrong responses attractors because, if they are well chosen, they will reflect the errors and misunderstanding that commonly occur. Difficult though it is, constructing questions is an illuminating activity for teachers but one that cannot be hurried. Attractors have to be collected and recorded over a long period of time. Almost every time we say 'No' to a pupil response (or its softer equivalents like 'Not quite' or 'Would you like to think about that again?') there is probably an attractor lurking. Teachers in training often take part in 'question and answer' sessions and experienced teachers might like to recall these to freshen up their own techniques. It is worth remembering that pupils seldom give either random guesses or answers which seem wrong to their way of thinking though, admittedly, some of them may be very tentative. It is a satisfying feeling when we hear right answers but it is much more useful to class learning if we concentrate on the wrong ones. Attitudes to the work are involved. If a student has made his offering and it is brushed aside with what may be only a hint of a frown and the teacher then goes on until a correct answer appears, which is greeted with however so slight a warm smile, then the atmosphere for most is not very encouraging. Beginning teachers find reactions to these responses very difficult because they do need a lot of experience. Their tutors often detect the very reasonable reaction: 'How on earth did he get that? Can't imagine! Pass on quickly'. It is a place where supervisors can be really helpful by encouraging the student to note these incidents when observing classes and to raise them in discussion with the teacher afterwards. The roles can be reversed when a teacher is observing the student at work. Discussion and activities on questioning can be found in *Mathematics Teacher Education Project* (1980) TS3, TS4, TS6.

Bearing in mind the importance of the wrong answers, multiple-choice questions can help considerably because we have an immediate count of the relative popularity of the various attractors. Any class we meet for the first time, after the infants, will have had very varied backgrounds and experience in their maths. Discrete diagnostic testing allows the teacher to choose which are the urgent matters for attention. It is unprofitable and not really sensible to blame deficiencies on the previous stage. It is a cyclic activity. Teacher trainers blame the undergraduate courses and so on down the chain; but they themselves were probably teaching in school or college at the beginning of the chain! In any case, part of the power of mathematics lies in the flexibility of having various transformations available to turn a task into an equivalent but easier one, which results in various methods of solution. For example, solving an equation usually involves transforming it into a different and equivalent but easier one, and there will be several possibilities.

Images and consequences

The reason for the persistence of the 'core of difficulty tasks' is not fully understood. One feature of them is that, on the whole, teachers do not reliably estimate the extent of the difficulties. With some tasks they are quite good at predicting the success of their students but with core of difficulty tasks they habitually overestimate. This may imply that specific attention paid to these topics could produce a cure. This does happen to a certain extent and a short Schools Council-funded project produced some evidence to that effect (Barr & Rees 1981). There is, however, the possibility that there really is some inherent difficulty in these tasks and they may be indicators of a degree of mathematical ability or reveal something about the structure of mathematical abilities. This side of it is discussed elsewhere in detail (Furneaux & Rees 1976, 1978; Rees 1981; Curnyn 1979).

One model of thinking which might account for the different order of difficulty is this: tasks in the core may all be ones in which several different images or intuitions are available. Corresponding to the particular image chosen there will be a routine or a calculation which may well be harder or more conceptually troublesome than that for a different image. In the activities of this chapter are some where small groups of teachers are asked to solve an elementary problem on their own and then asked to talk their way through their various solutions. Most are surprised by the number of different starting points. This aspect of the difficulties accounts for our backing up the multiple-choice questions with tape recordings or other 'talk-through' methods.

An example, adapted from one which occurred with us, is this: Find x if $\frac{4}{15} = \frac{x}{5}$. An apprentice started, 'To get the 5 at the bottom I must divide 15 by 3. So I have to divide the top by 3. 3's into 4.... You can't! I can't do this one I am afraid.' He sees the problem as one of equivalent fractions. He could actually divide 4 by 3 in some contexts. If asked to divide a rod 4 m long into three equal parts he could have done it. He might not have been able to divide four cakes equally between three people but then that is not something that we often do. He could use ratio ideas reasonably well in some contexts but having come to the view that he was dealing with fractions the idea of a top that was not a natural number bothered him, quite properly and reasonably. He is revealing that he does know quite a lot; but he is also revealing where there is a difficulty. If we are thinking of marking on a right or wrong basis we could only give him 0 for his efforts. He might have chanced his arm and written 4/3 (which must have scored a mark) but he is reluctant to think of it as a *number* at all. And yet 4 divided by 3 really is 4/3. Writing it as a mixed number or a recurring decimal fraction is merely a transformation to a different form, not finding an answer.

It is not surprising that work with fractions causes persistent difficulty. One of the activities we suggest sends teachers to textbooks to analyse what they really say about fractions and there are some very curious statements to be found. The trap, however, is that two or three statements in a book may separately look sound but be mutually inconsistent. Fraction difficulties are not avoided nevertheless by claims that the calculator, decimal notation and the metric system make it unnecessary to bother with fractions. There are many situations where the fraction concept is needed, especially where it blurs into a ratio concept and even more so as it grows into a concept of algebraic fractions. For a beginner, problems concerned with 17 teeth on one gear wheel meshing with 5 on another are not very much helped by decimal notation, even though the engineer may express his gear ratios that way. The gear ratio example is interesting. A ratio expressed as 1.63 : 1 gives us useful information but it does not tell us how many teeth to cut on the wheels. In case it is thought that gear wheels might be left to technology classes and apprentices it is worth adding that gears (and pulleys) give the only really physical modelling of multiplication of one directed number by another. Most book explanations cheat by providing models in which a directed number is really multiplied by a scalar. [For a good exposition of this see *Mathematics Teacher Education Project* (1980) CT6.]

Textbooks often suggest that we 'simplify' a set of expressions in an exercise but the question to ask in return is 'for what purpose?' Very often the expressions are already in the simplest form for *some* purpose. Examiners have been known to suffer from this defect. They will only give full marks for their version of the simplest form which mirrors the most general question teachers ask: 'guess what is in my mind!' Is it, for example, obvious that a fraction has to be expressed 'in its lowest terms' to be in its simplest form for all purposes? Instructions for marking in a SLAPONS booklet generally allow equivalent fractions as accepted answers but then insist that top-heavy fractions must be converted into mixed numbers to be accepted. Is it obvious that $1\frac{1}{16}$ is necessarily a simpler form than 17/16? (SLAPONS 1980).

The answer is 'it all depends'. What it depends on is how a student perceives a problem. There are interesting comments which follow this up in CSMS (1981) where there is a description of childrens' tendency to invent almost any addition strategy, however complicated, to avoid multiplying by a fraction. Comparisons of mathematics learning with learning music or sport are often valid: the simple elegance of the professional represents extreme sophistication and comes after hard practice.

Routines and imagination

Among the discussions arising from this work will be many on the extent to which mathematics has to consist of routines and mechanical work. The answer has to be some sort of compromise because the extremes are fairly obvious: solely mechanical methods can stifle imagination and discourage flexibility but free invention and investigation can be self-indulgent and lead to rediscovering the wheel too often. The right mixture probably depends most on characteristics of the learner but must include characteristics of the teacher as well. We do not really know enough about this yet. Positive attitudes and a reasonable feeling of success, with the right amount of challenge, are what we have to aim for. One piece of self-criticism we can all carry out from time-to-time is to reflect on whether our classes are getting 'hooked' on us as people or on our methods and therefore not seeing enough variety to be guided to independence.

It is interesting that the microcomputer is encouraging revived attention on algorithmic working, not in the sense of dead rituals and pencil and paper methods but in allowing the student to construct and test short routines that will stand the demands made on them. Sets of good routines can then be marshalled for more imaginative problem solving. This is not different from the best teaching of the past but the processes are both made more explicit and rendered more powerful.

Context

Questions of context have been raised in many test items so far. It is frequently said that if maths is made relevant to some everyday practical activity then students will find much less difficulty with it. True in parts. Employers and training officers often berate the schools because their recruits cannot perform certain tasks on entry and within a few weeks are using them fluently in their work. However, if only 10% of our pupils enter the industry in question, is the activity practical and relevant for the others? The answer is, partly. The pay-off for many of our more formal techniques is simply that they are available for many contexts. We are left to compromise our way through the dilemma: shall we teach the technique and learn to apply it or tackle several practical activities and establish the technique as a way of dealing with all the problems in common? Sociologists will tell us that this remote kind of motivation will not have much sense of urgency for most children.

A friend, an outstanding musician, was conscripted into a military band. One week he was posted to play the harp at a concert. 'But Serge! I can't play the harp'. Response: 'Well! Yer've got till Thursday ain'tcher'. Perhaps that would be a suitable inscription over the doors of our schools if they are to relate more closely to the world of work.

Two more quotations may prompt serious discussions in groups of teachers. A 20-year-old woman was being questioned before starting a numeracy course. Can you divide 6 by $\frac{1}{2}$? Answer...3. How many halves in 6? Answer...3 What is 6 over $\frac{1}{2}$? Answer...3. How many $\frac{1}{2}$ps are there in 6p? Answer... Oh! 12, of course. I didn't know you were talking about real things like money'.

A 7-year-old was doing a sum:

$$\begin{array}{r}31\\-1\underline{2}\\\hline\end{array}$$

He starts: '1 from 3 is 2; 2 from 1... you can't.' Then he was asked: 'There were 31 people on a bus and 12 got off; how many were left?' Almost instant answer (in his head)...'19'.

The second quotation is from Schools Council Working Paper 72 (Low Attainers in Mathematics 5–16 1982) which contains many other good examples.

Are these really cases of 'formal operations' beyond the students' capabilities? Would some structural apparatus like Dienes' Multibase Arithmetic Blocks help the 7-year-old? If it did, would he make the connection with the people on the bus where some form of 'counting-on' would be appropriate? How can we help students to see mathematics as offering a real advantage in tackling many different problems with one technique; one problem instead of several?

We have shown question items in a pure form with corresponding context forms. The number of correct responses changes but an interesting question is: do the same wrong responses maintain their popularity in the different contexts? We are not sure. Others may like to investigate.

Activities for teacher training

The following activities are suggested for training sessions. Most of them require some classroom experience though beginners may find some of them helpful. As teachers we need to protect ourselves from too much guilt over the errors we find in childrens' work. Reports over the last 100 years have complained about the number of errors but there are plenty of measures to suggest that there has been much improvement. The main thing to watch may be over-enthusiasm in developing the 'too slick' kinds of rule that buy spurious short term gains. The difficulties we are discussing occur worldwide. They are least troublesome in places where learning is very formal and authoritarian, for example in east European and Chinese communities but the demands of our education have been more complex and less competitive than in these countries and carry a hope of more individual development and autonomy. Rote learning is, of course, possible and occasionally may be necessary. The pencil and paper methods of finding square roots and highest common factors are examples. It is possible to defend the opening steps of the square root algorithm, even now, because it gives a good approximation with the decimal point fixed in the right place for very little effort with even the most awkward numbers. [If there are no veterans in the group to remember how to do them they can be found on p.10 and 274 of Durell's (1936) *General Arithmetic for Schools*.] Very few of us could produce good reasons for the methods. Perhaps children do not demand good reasons, just reasonableness. Even that may be too idealistic. A very happy child came home to announce that she had been doing 'line sums' at school. 'Oh good', we said, 'er... what are line sums?' 'Well. You put down a 3 and a line under it, and a 4 under that and then an add sign. Then you put 2 and a line and a 5 under it; and then you put a long line and say four 5s are 20...'. She was happy because she had got them all right. Perhaps we should make rather heavier demands than that?

Group activities

The activities that follow consist largely of suggestions for discussion with occasional collection of examples. Comments added should be taken as tendentious rather than fixed truths.

Exercise 1

It is interesting to reflect on our own working of elementary problems and to record very quickly what our approaches to the problem have been. It needs an honest response to the task if we are to record the whole process, including the false starts. A very elegant and condensed solution is not always the first that comes to us. Almost any of the exercises in this book and dozens of similar ones from standard texts can be used. Offer them one at a time to the group as very brief, say 20 second, tasks and get members to declare their solution methods. Groups are usually surprised at the sheer variety of method produced and the differences that can occur in individuals when a particular operation is required in different contexts. It is a fairly safe indicator that an item for which there is not much variety of response is not a good diagnostic item and that one with a big variety is a good diagnostic item belonging to the core of difficulty. Inservice teachers can take the items discussed back to their own classes together with their forecasts of which will turn out to be popular attractors and of what fraction of their classes will get the items right. Student teachers can often negotiate conditions for trying similar work but, as usual, testing of any kind does require delicate negotiation and clear explanations.

Exercise 2

Constructing multiple-choice questions is quite difficult and is a suggested activity in Exercise 7. A useful preliminary is to consider items in this book to discover the origins of the misconcepts behind the various attractors. An attractor which attracts no response is useless and all those used here have, in fact, attracted some response, however perverse some of them may seem. For example:

Task 1. 0.3 x 0.3 is?

a	0.09	Key
b	0.6	addition or 'three 3s are 6'
c	0.9	decimal point alignment or 'three 3s are 9 so put a point in front'
d	0.06	'3 twice is 6 and there are two decimal places'

2. How much VAT at 15% would be charged on an article costing £12?

a	15p	15% means 15p
b	18p	1% plus half of that

c	72p	1% plus 5%
d	£1.26	10% plus 0.5%
e	£1.80	Key

Some of these seem absurd but a great deal of testing shows that the various attractors and the key do occur with fairly predictable frequencies. For the VAT question we have had responses a, b and d from 10 to 20% of the students tested, for c 15 to 30% and for the key d from 40 to 60%.

Exercise 3

A school is reporting on some diagnostic tests given to children on transfer from primary to secondary school:

'It was decided, from the outset, that computational skills and mathematical concepts would be tested separately in two distinct tests.'

(The teachers all approved of this but a tutor who had taught them on a course in assessment and a psychologist, introduced to monitor and assess the results, were not so happy.)

[From *Interface* (1983) Mathematical Association]

Examine the questions at the end of any chapter in the book and decide, if possible, a classification into 'computational skills' and 'mathematical concepts'. Are there any items which belong in neither class? Or both? Is there any way these terms could be defined to make a useful distinction?

Exercise 4

Is it worth bothering with fractions in the days of metric units, the calculator and the home computer? We think it is because the 'fraction concept' is fundamentally important and growing points for it can be provided at early levels. Nethertheless, difficulties are understandable. Generalizing the multiplication rule as $(a,b) \times (c,d) = (ac,bd)$ and the addition rule as $(a,b) + (c,d) = (ad + bc, bd)$ shows what we are really asking, even though we do not ask for it to be expressed in that way.

Physical models for adding fractions are fairly obvious. It is worth pointing out that adding and multiplying *numbers* hides the fact that in the real world adding fractions only makes sense if they are measures of the same kind. $\frac{1}{2} + \frac{1}{3}$ must refer to $\frac{1}{2}$ and $\frac{1}{3}$ of the same thing (or equivalent things) with an answer in the same units. This is not true of multiplication where the $\frac{1}{2}$ and $\frac{1}{3}$ can be of different units, for example kg and cost per kg. A useful discipline in all fraction and percentage work is to add 'of what' every time either is used. If I take an exam I may say, 'I got 60%. I might say, instead, 'I got 3/5' but I would not unless I added 'of the possible marks'. Try this in group work. Search for textbook examples and insert the 'of what' every time. We would not want to do it always but it is important to be able to do it. There is an important clash here between fraction as a pure number mapped on to a number line and fraction in its elementary sense of part of a whole.

It is possible to model the multiplication of fractions by using the simplest intuitive view of area rather than the dimensionally complex 'length × breadth'. Is it hopeless to encourage writing, for the area of a rectangle 5m by 3m, $A = (5 \times 3)$ m^2 rather than 5m x 3m?

Start with a rectangular piece of paper (A4 size works well).

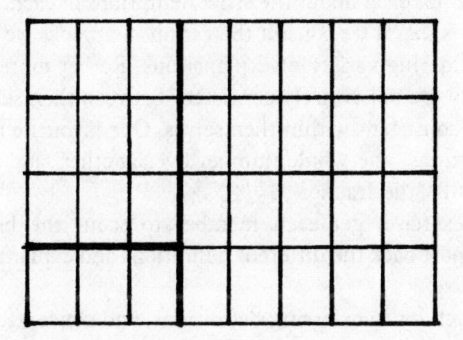

Fold it into eight equal pieces one way. The intuitive feeling of eight equal parts does not need any subtle notion of area. 3/8 of the whole is also obvious. Now fold into four the other way. 3/4 of 3/8 is then clearly marked. The transition from 'of' to 'multiplied by' is often brushed over but it is plain here because our target piece has 'three rows of three' just as in multiplying natural numbers. In the same way the '4 x 8' is how many parts the whole has been divided into. The required piece has nine of them. Notice the unit distances along and across are not the same. It is easy to answer the questions '3/8 of what?' and '3/4 of what?'. The answer involving '9/32 of what?' is also clear. Later we can deduce the 'length x breadth' ideas when we are to use standard units for the lengths and standard *squares* derived from them for the area. But that is a long way ahead. The three rows of three can, of course, be scattered in other ways over the grid.

Try this modelling for other fraction multiplication. We have modelled binary operation on a two-dimensional drawing. Try a folding technique with all the folds parallel. It is possible but the division into 32 basic units is by no means obvious and ignores the 'row by column' feeling normally acquired with natural numbers. Modelling is important but some models communicate more readily than others. Unfortunately there is no comprehensive model which communicates well all the ideas of number relations we want.

Fractions as 'part of a whole' are often illustrated by the circular cake. Is there any folding technique to '1/2 of 3/4 of a cake'? What other shapes can the folding technique be extended to? Triangle? Parallelogram?

Exercise 5

The main trouble with fraction, percentage, ratio and proportion is that most of us find we are not clear about the strict definitions of each. The reason why we are not clear is that if we consult the various textbooks we normally use we will find a bewildering variety of explanations. Rather more surprising than the fact that they are not entirely consistent between themselves is that some of them are not consistent within themselves. Our favourite is one that says a fraction is written as 'one whole number over another' and half a page later asks us to simplify 'the fraction $3\frac{1}{2}/4\frac{1}{3}$'.

As a group exercise, get each member to scour the books he/she has available and report back the different definitions and explanations for each of the four concepts.

We cannot teach by transferring definitions, and concepts, in any case, are defined partly by their use but we do need to be more clear and consistent about these matters. It is quite a good idea to look at the books used by our grandfathers (and the occasional grandmother). They are often superbly accurate. The fact that they are pitched in stiff Victorian language and as a result sound ponderous and pedantic should not cloud their virtue where they have it.

Hall & Knight's *Higher Algebra* defines *ratio* as 'the relation which one quantity bears to another of the *same* kind, the comparison being made by considering what multiple, part or parts, one quantity is of the other'. That is exactly right and suitably modern because it really focusses on the scale-factor aspect. If the ratio of the number of boys to the number of girls in a school is 2:3 the only useful deduction is that the number of boys is $\frac{2}{3}$ times the number of girls and the inverse result that the number of girls is $\frac{3}{2}$ times the number of boys. The ratio itself simply says that there $1\frac{1}{2}$ girls per boy or 2/3 of a boy per girl which is of doubtful application.

Single ratios on their own are seldom of interest. Sets of equal ratios, however, constitute proportion which is exceptionally important. It represents a whole range of 'multiplicative structures' as against additive ones. The equal ratios are not always apparent in ordinary usage. For example, when we say that a concrete mix has cement, sand and ballast in the proportions 1:2:3, what are the equal ratios we can form?

The pure ratio definition requires the two quantities concerned to be of the same kind and in the same measure. That one of them is k times as much as the other then makes sense. This causes trouble in science teaching. A physicist may have a law about gases which says the volume is proportional to the temperature. The strict 'equal ratios' approach demands that we write

$$\frac{V_1}{V_2} = \frac{T_1}{T_2}$$

and we have the ratios as pure numbers. The physicist, however, will want to write

$$\frac{V}{T} = k, \text{ leading to}$$

$$\frac{V_1}{T_1} = \frac{V_2}{T_2}$$

which is much more difficult intuitively unless we are careful enough to make clear it is a statement connecting magnitudes and not measures. It is not sensible to say the volume is k times the temperature unless units are clearly attached to k. It has then become a quantity rather than a ratio. There is no resolving this dilemma because current practice with SI units leaves it understood that a symbol for a quantity includes its units. Scientists should realize how difficult this is and that some explicit work is needed to establish that, for example,

$$\frac{V_1}{V_2} = \frac{T_1}{T_2} \quad \text{implies that} \quad \frac{V_1}{T_1} = \frac{V_2}{T_2}$$

Inservice groups should raise these questions with their science colleages if they have not already done so and discuss their findings and conclusions with the rest of the group. We do know that there is a very great difference in success with items expressed in a way which leads naturally to the first form (starting with the ratio of the volumes) and with those which lead to the second.

Another topic worth discussion at this point is that of the advantages to be gained by treating the trigonometric ratios in the way presented in the first SMP textbooks. The sine and cosine are defined as lengths projected on the axes as a unit line rotates and the ratio idea is transferred to enlargement from a standard triangle for the usual side and angle calculations. The language used there is of the 'unit rotating vector' but that can easily be avoided if required.

Exercise 6

The errors arising in the statistics section illustrate a useful teaching point. It shows most obviously in questions on the median. Students often remember that it is 'the middle of something' but settle for the wrong 'something'. The word median used like this is an abstract noun. Its abstract status can be lowered for beginners by using it always as an adjective, for example in the form 'median score', 'median length' or 'median shoe-size'. This fits in with the median as a representative of the scores, lengths or shoe-sizes and helps in moving toward the right 'something'. Are there any other ideas for which the level of abstraction is lowered by this careful use of language?

For those still trying to work from set language, 'domain' and 'domain set' form one example. 'Common fraction' and 'decimal fraction' form another. Student teachers often start a lesson with a question such as, 'Can anyone tell me what a decimal is?'. That is very difficult to answer without a long explanation. One level down in difficulty is to answer the question, 'What sort of fraction is a decimal fraction?'

Exercise 7

The time has come to write your own test items. Writing good ones is difficult and needs practice but we have to start somewhere. It is usual to advise testing one idea only in each item. The trouble with that is that the students' perception of one idea may not be the same as ours. Still, it remains sensible to keep the stem of the question as simple and unambiguous as possible. Then the difficulty of the question is fixed by the subtlety of the differences in the answers offered. As a start, try writing questions which run closely parallel to those in this book. The results should then be exposed to the rest of the group for comment and criticism.

The next step is to devise items from topics in your current teaching where the things that go wrong may be in your mind. A good long term plan is simply to keep a file of recurring errors discovered when marking or supervising work. Questions can then be used for quick classroom tests or revision with flash cards as well as for formal diagnostic testing.

References

Barr G.V. & Rees R.M. (1981) The use of diagnostic assessment in the classroom: an exploratory study. Brunel University (MEG) Report for the Schools Council.

CSMS (1981) *Children's Understanding of Mathematics* 11-16 (K.M. Hart ed.) London: John Murray.

Curnyn J. (1979) The Structure of Mathematical Ability. Proceedings of the Third International Conference of IGPME, University of Warwick.

Durell C.V. (1936) (and many later edns) *General Arithmetic for Schools* London: Bell.

Examinations and Assessment (1968) *Mathematics Teaching Pamphlet* No. 14 Association of Teachers of Mathematics.

Furneaux W.D. & Rees R.M. (1976) *Occassional Publications Series* No.1 Brunel University, Dpt Education.

Furneaux W.D. & Rees R.M. (1978) The Structure of Mathematical Ability, *British Journal of Psychology* **69,** 507-512.

Interface (1983) Leicester: Mathematical Association.

Low Attainers in Mathematics 5-16 (1982) *Schools Council Working Paper 72* London: Methuen Educational.

Mathematics Teacher Education Project (1980) London: Blackie.

Rees R. (1981) Mathematically Gifted Pupils: Some Findings From Exploratory Studies of Mathematical Abilities *Mathematics in School* (The Mathematical Association) Vol. 10, No.3.

SLAPONS (1980) *Information Booklet 1980-1981 for School Leaver's Attainment Profile of Numerical Skills.* (Now administered by the Royal Society of Arts.)

Tests (1978) Leicester: Mathematical Association.

A FINAL COMMENT

'Mathematics is the Language in which God has written the Universe.'
(Galileo)

'Mathematics provides a means of communication which is powerful, concise and unambiguous.'
(Cockcroft Report)

These two quotations are separated by many centuries and yet both express the essential nature of mathematics ... its power to communicate. But the effectiveness of this communication lies in our hands and we are all only too aware that communication is one of the most difficult and vunerable of human activities! The quality of our students' learning must surely reflect the quality of our teaching.

The central message of this book is the need for us as teachers to both effectively diagnose and prescribe. In so doing, however, we must beware of falling into the trap of being so demoralized by our students' difficulties that we underestimate our students' potential. Awareness of the nature of mathematics and the misconcepts which can arise is the major step towards effective prescription.

How then should we teach? Inferential, insightful approaches should be fostered in our students as early and as consistently as possible for the sake of their attitudinal as well as mathematical development. Let's therefore encourage our students to *explore* maths, to develop a divergent thinking approach to problem solving as well as the more traditional convergent thinking approach. But of course we must always remember the need eventually to cement our students' understanding by the establishment and practice of mathematical techniques.

Throughout our teaching we should be aware of our use of mathematical language. We may do our students a disservice by using language which is mathematically imprecise. Ambiguity can create the very confusion we try so hard to avoid. The power of mathematics lies in the precise, concise, language in which generality is expressed. This language enables us to describe the many aspects of the world around us as Galileo stated so many years ago.

We should aim to help *all* our students, not just a chosen few, to an appreciation of the power of this language by helping them to develop a feeling for number and space, more generalized symbols and relations. And this exposure will stimulate some of our students but, as we know only too well, will also frighten many more, and for these perhaps, a more 'softly, softly' approach will be needed. There is too much failure associated with maths. We must aim for confidence building and this means generating *success* for our students. And this implies effective diagnosis, effective prescription and effective setting of appropriate tasks which will generate that confidence and success.

Look, Listen and Learn with your students. Be Aware, Explicit and Flexible so that they acquire Concepts with Competence and Confidence. Then the concepts acquired by the confident 13-year-old may well be the crystallized concepts of the confident adult.

DIAGNOSTIC ASSESSMENT

Read the following carefully:
1. Try to answer all questions and work as carefully as possible.
2. Each question shows four possible answers a,b,c,d one of which *may* be right. Decide which you think is correct then put a ring around it. If you think none of the answers are correct put your own answer in e.

Try the two examples:

Task 80 x 4 is?
    a 20   b 84   c 320   d 32   e
If you think the answer is 320 ring c thus: ⓒ

Task 45 + 15 is?
    a 50   b 3   c 30   d 675   e
If you wish to put an answer of 60 in e, it should be written thus:
    a   b   c   d   ⓔ **60**
If you make a mistake, cross it out and put a new ring thus:
    a   ⓑ   c   ⊗   e

Diagnostic assessment

1. 967 + 2056 is?
    a 3023   b 2923   c 2913   d 2023   e

2. 904 − 586 is?
    a 418   b 422   c 428   d 482   e

3. If $\frac{2}{x} = \frac{1}{5}$ then x is equal to?
    a 10   b 5   c $\frac{5}{2}$ d $\frac{2}{5}$ e

4. If the sides of two squares are in the ratio 1:5 the ratio of their areas is?
    a 1:5   b 1:6   c 1:10   d 1:25   e

5. A friend intends to holiday in the United States. He wants to take £70 to spend. If the exchange rate is $1.62 to the pound, the number of dollars he will receive for his £70 is?
    a $11340   b $1134   c $113.4   d $11.34   e

6. Calculate $32\overline{)6432}$
   a 21   b 201   c 210   d 2010   e

7. A foreman reports that 75% of the work force arrive at least 10 minutes early after lunch. If there are 40 workers altogether, how many arrive at least 10 minutes early after lunch?
   a 10   b 10%   c 30   d 30%   e

8. 0.2 x 0.3 is?
   a 6   b 0.6   c 0.5   d 0.06   e

9. It was reported after a survey of Dover to Ostend car ferry users, that half were on their way to a destination in Belgium, a third in Holland and the rest in Germany. What fraction of those surveyed were on their way to Germany?
   a $\frac{1}{6}$ b $\frac{1}{5}$ c $\frac{1}{3}$ d $\frac{2}{5}$ e

10. 5.21 − (3.69 + 1.15) is?
    a + 0.37   b + 1.37   c + 2.67   d − 0.37   e

11. The average of 10, 8, 9, 11 is?
    a 38   b 11   c $9\frac{1}{2}$ d 8   e

12. 0.8 ÷ 0.8 is?
    a 0   b 0.1   c 1   d 8   e

13. A report contains 31 pages. If I need to photocopy the report 30 times how many sheets of paper will I use?
    a 962   b 961   c 931   d 930   e

14. $\frac{9}{64} + \frac{5}{16}$ is?
    a $\frac{45}{80}$ b $\frac{29}{64}$ c $\frac{14}{64}$ d $\frac{14}{80}$ e

15. If $p = 4$ and $q = 5$ then p^2q is?
    a 400   b 80   c 40   d 21   e

16. On a map of scale 1 to 50 000 the distance between two ancient sites is represented by a line 10 mm long. The actual distance between the ancient sites is?
    a 5 m   b 50 m   c 500 m   d 5000 m   e

17. 18 components out of a bath of 360 were scrapped: as a percentage this was?
    a 2   b 5   c 20   d 50

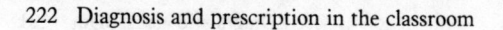

18. $\frac{7}{8}$ written as a decimal fraction is?
    a 7.8   b 1.1   c 0.78   d 0.875   e

19. If the diameters of two circles are in the ratio 1:5 the ratio of their areas is?
    a π^2:5   b 1:5   c 1:10   d 1:125   e

20. A jogger takes the same route on each of 7 days. The times required to complete this route are noted as 16, 19, 16, 20, 22, 17 and 16 minutes. The median time taken in minutes to complete the route is?
    a 16   b 17   c 18   d 20   e

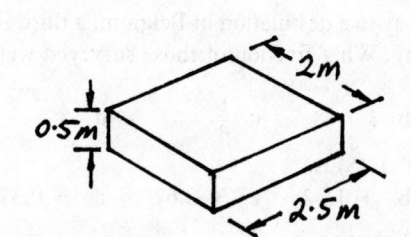

21. The volume of solid shown in the figure is?
    a 250 m³   b 25 m³   c 5 m³   d 2.5 m³   e

22. A bill adds up to £11 without VAT. When VAT at 15% is added to this total, the amount to be paid is?
    a £11.15   b £11.66   c £12.16   d £12.65   e

23. A circular lampshade has a diameter of 10 cm. The length of cotton fringe needed to go once around the lampshade, if π is taken as 3.14, is?
    a 314 cm   b 62.8 cm   31.4 cm   d $(3.14)^2$ x 10 cm   e

24. A cocktail is made up in the following proportions
    1 part brandy
    6 parts white wine
    3 parts orange juice
    The amount of orange juice needed to make 500 ml of the cocktail is?
    a 430 ml   b 300 ml   c 150 ml   d $83\frac{1}{3}$ ml   e

25. $4\frac{5}{16} - 2\frac{3}{8}$ is?
    a $1\frac{1}{16}$ b $1\frac{15}{16}$ c $2\frac{2}{8}$ d $2\frac{2}{16}$ e

26. The circumference of a circle diameter d is?
    a $2\pi d$ b πd^2 c $\pi^2 d$ d πd e

27. If $2x + 7 = 15$ then x is equal to?
    a 4   b 6   c 8   d 11   e

28. 0.7 x 0.7 is?
    a 49   b 4.9   c 0.49   d 0.049   e

29. A score of 60 marks out of 75 is needed to pass a practical test. This score as a percentage would be?
    a 90%   b 80%   c 60%   d 45%   e

30. 1.62 x 70 is?
    a 11.34   b 113.4   c 1134   d 11340   e

31. If $\frac{1}{x} = \frac{7}{8}$ then x is?
    a $\frac{7}{8}$ b 7   c 8   d $\frac{8}{7}$ e

32. 75% of 40 is?
    a 35   b 30   c 20   d 4.8   e

33. A piece of wire is 5.21 metres long. If it is to be used to make two extension leads, one 3.69 metres long and the other 1.15 metres long, then the wire will be?
    a 0.37 m too short
    b 0.53 m too long
    c 1.37 m too long
    d 0.37 m too long
    e

34. $1 - (\frac{1}{2} + \frac{1}{3})$
    a $\frac{5}{6}$ b $\frac{3}{5}$ c $\frac{2}{5}$ d $\frac{1}{6}$ e

35. The number of hours of sunshine at a seaside resort over 4 days was reported to be 10, 8, 9 and 11 hours. The average number of hours sunshine is?
    a 8   b $9\frac{1}{2}$ c 11   d 38   e

36. 31 x 30 is?
    a 930   b 931   c 961   d 962   e

37. The area of a circle of radius r is?
    a $2\pi r$ b πr^2 c $(\pi r)^2$ d $\pi^2 r$ e

38. If $p = 4$ and $q = 5$ then pq^2 is?
    a 400   b 100   c 40   d 29   e

39. If the sides of two cubes are in the ratio 2:3 the ratio of their volumes is?
   a 2:3   b 4:6   c 4:9   d 8:27   e

40. During the last few sessions of a course the number of trainees at a 2.00 pm lecture on Friday were as follows: 14 15 13 13 0 5. The average number of trainees per session was?
   a 60   b 12   c 10   d 8   e

41. 783 ÷ 27 is?
   a 281   b 30   c 29   d 28   e

42. With π = 3.14, the area in mm^2 of a circle of diameter 20 mm is?
   a 31.4   b 62.8   c 314   d (3.14)^2 x 20   e

43. If $m = 3$, $x = 0$ and $k = 5$ the value of y, if $y = mx + k$, is?
   a 15   b 8   c 5   d 0   e

44. If the diameters of two spheres are in the ratio 2:1 the ratio of their volumes is?
   a 2:1   b 4:1   c 8:1   d $2:\pi^3$ e

45. A sales assistant notes that on average, tour guides are sold more than any other books in his department. This type of average is the?
   a mean   b median   c middle value   d mode   e

46. In the number 345 021 the value of the four is?
   a 40 000   b 4000   c 400   d 40   e

47. A cocktail is made by mixing 100 ml of vermouth and 150 ml of gin. The vermouth and gin are in the ratio?
   a 150:100   b 250:1   c 1:25   d 2:3   e

48. The mean price per pound of potatoes over a 5-week period was 7p. If the price goes up this week to 10p a pound what is the mean price per pound over the 6-week period?
   a 10p   b $8\frac{1}{2}$p   c $7\frac{1}{2}$p   d 3p   e

49. The value of 278 956 is approximately?
   a thirty thousand
   b three hundred thousand
   c two million eight hundred thousand
   d thirty milion
   e

50. $5\frac{1}{2}$ can be written as?
   a $\frac{51}{2}$ b $\frac{11}{2}$ c $\frac{10}{2}$ d $\frac{6}{2}$ e

SUGGESTIONS FOR READING

Government publications

APU reports

Mathematical Development, Assessment of Performance Unit.

Primary Survey Report No. 1 (1980) HMSO.
Primary Survey Report No. 2 (1981) HMSO.
Primary Survey Report No. 3 (1982) HMSO.
Secondary Survey Report No. 1 (1980) HMSO.
Secondary Survey Report No. 2 (1981) HMSO.
Secondary Survey Report No. 3 (1982) HMSO.

HMI reports

Curriculum 11–16 (1977) HMI Discussion Documents. DES: HMSO.
Primary Education in England (1978) A survey by HM Inspectors of Schools. DES: HMSO.
Mathematics 5–11 (1979) A handbook of suggestions. Matters for discussion. DES: HMSO.
Aspects of Secondary Education in England (1979) A survey by HM Inspectors of Schools. DES: HMSO. A further booklet is available: *Supplementary Information on Mathematics*.
Mathematics Counts (1982) Report of the Committee of Inquiry into the Teaching of Mathematics in Schools under the Chairmanship of Dr W.H. Cockcroft.

Books and articles

Bolt B. (1981) *A Resource Book for Teachers: Mathematical Activities* Cambridge University Press.
Burkhardt H. (1981) *The Real World of Mathematics* Blackie. (Has a great variety of 'context' questions.)
Buxton L. (1981) *Do you Panic about Maths? Coping with Maths Anxiety* Heinemann Educational.
Dienes Z.P. (1971) *Building up Mathematics* Hutchinson Educational.

Hart K. (Ed.) (1981) *Children's Understanding of Mathematics 11–16.* John Murray.

Howson G. & McLone R. (1983) *Maths at Work* Heinemann Educational.

Skemp R.R. (1976) *Relational understanding and instrumental understanding.* Mathematics teaching. *Journal of the Association of Teachers of Mathematics* No. 77.

Skemp R.R. (1971) *The Psychology of Learning Mathematics* Pelican Books. [Also: (1979) *Intelligence, Learning and Action* Wiley.]

Thompson D'Arcy (1961) *On Growth and Form* (abridged edn). Cambridge University Press.

Brunel University, Department of Education
Mathematics Education Group
Research Projects and Reports

1. Diagnosis of mathematical difficulties

Department of Education and Science

Further education
Difficulties in Mathematics Experienced by Craft and Technician Students September 1971 (Rees).

University
Difficulties in Mathematics Experienced by Brunel First-Year Engineering Students December 1971 (Rees).

Secondary school
Difficulties in Mathematics Experienced by Secondary School Students July 1973 (Rees).

Teacher training
Mathematics in Colleges of Education : Some Difficulties Experienced by Teachers in Training July 1973 (Rees).

2. Diagnosis and structure of abilities

British Petroleum

Secondary school
Learning Difficulties in Mathematics. A Further Investigation of the Nature of the Determinants of Mathematical Performance July 1978 (Rees & Furneaux).

Nuffield Foundation

Further education
Some Common Mathematical Difficulties Experienced by Technician Students at TEC levels I and II November 1979, December 1980 (Barr & Rees).

3. Diagnosis, structure of abilities and affective factors

Leverhulme Trust

Primary and early secondary schools
An Experimental and Statistical Study of the Structure and Development of Mathematical Ability during the late Primary and Early Secondary Stages of Education, together with an Investigation of some of its Cognitive and Affective Concomitants September 1980 (Furneaux, Curnyn).

Articles

Barr G. (1980) Some student ideas on the median and the mode. *Teaching Statistics* **2** (May).

Barr G. (1981) *An Overview of an Exploratory Study: The Use of Diagnostic Assessment in the Classroom* Proceedings of the BSPLM Conference, Bath University, September 18-20.

Barr G. (1983) The operation of division and the embedded zero. *Mathematics in School* (The Mathematical Association) **12**, No. 4, September.

French P. & Barr G. (1983) *364 Adds Up to Confidence: The Development of the City and Guilds Scheme in Numeracy* TES, 25 March.

Furneaux W.D. & Rees R. (1976) *The Dimensions of Mathematical Difficulties* A set of problems proving more difficult than teachers expect at all levels in the educational system. Brunel University, Department of Education. Occasional Publications Series No. 1, September.

Furneaux W.D. & Rees R. (1978) The structure of mathematical ability *British Journal of Psychology* **69**, 507-512.

Rees R. (1974) An investigation of some common mathematical difficulties experienced by students. *Mathematics in School* (The Mathematical Association) **3**, No. 1, January.

Rees R. (1974) *A Learning-Teaching-Learning Cycle?* TES Maths Supplement, April.

Rees R. (1975) *Brunel's New Language Laboratory Equation* TES Resources, December.

Rees R. (1976) *Mathematics in Teacher Training Institutions. Some Difficulties Experienced by Teachers in Training* IMA Bulletin, December.

Rees R. (1981) Mathematically gifted pupils: some findings from exploratory studies of mathematical abilities. *Mathematics in School,* May.

Books

Rees R. (1973) *Mathematics in Further Education. Difficulties Experienced by Craft and Technician Students* Hutchinson Educational.

Rees R. & Barr G. (1982) *Foundation Numeracy* Edward Arnold.

BBC radio continuing education

Rees R. (1983) *Maths with Meaning* Six programmes for maths teachers. Available as a package of three 60 minute cassettes plus booklet from Livingston Studios, Brook Green, London N22.

Note

Enquiries concerning these or other publications should be sent to the Secretary, Mathematics Education Group, Department of Education, Brunel University, Shoreditch Campus, Cooper's Hill, Egham, Surrey TW20 0JZ.